Secrets of the Night Sky

Secrets of the Night Sky

THE MOST AMAZING THINGS IN THE UNIVERSE YOU CAN SEE WITH THE NAKED EYE

Bob Berman

Illustrations by Alan McKnight

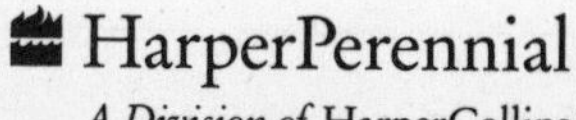

HarperPerennial
A Division of HarperCollins*Publishers*

This book was originally published in 1995 by William Morrow and Company, Inc. It is here reprinted by arrangement with William Morrow and Company.

HarperCollins books may be purchased for educational, business, or sales promotional use. For information please write: Special Markets Department, HarperCollins Publishers, Inc., 10 East 53rd Street, New York, NY 10022.

First HarperPerennial edition published 1996.

Designed by Brian Mulligan

Library of Congress Cataloging-in-Publication Data

Berman, Bob.
Secrets of the night sky : the most amazing things in the universe you can see with the naked eye / Bob Berman ; illustrations by Alan McKnight. — 1st ed.
p. cm.
Originally published: New York : W. Morrow, 1995.
Includes index.
ISBN 0-06-097687-X
1. Astronomy—Observers' manuals. I. Title
QB63.B473 1996
520—dc20 95-46182

96 97 98 99 00 RRD 10 9 8 7 6 5 4 3 2

Preface

Here on Earth we are enveloped in wonder—from the city within each cell to the power and mystery of a common thunderstorm. It is then no surprise that the universe beyond holds marvels beyond our dreams. This book explores those that belong to the night and to the sky.

Most of these astonishing phenomena lie at numbing distances and possess exotic properties. But many of the night's secrets begin right here! Mind-stretching explorations await us in our immediate neighborhood and envelop us as soon as the sun sets. Such "ordinary" experiences as twilight and blackness are, when we examine them, as astounding as the smoke rings of "forbidden radiation" that float among the night's distant suns.

We could have taken many paths in our journey of exploration. One road might have started with the familiar and led increasingly into the realm of the obscure. Or we could have begun with things close at hand and then proceeded to phenomena ever farther from Earth. Instead, we'll follow nature's lead, allowing the change of seasons to parade its wonders before us.

The book, thus divided into four sections corresponding to the seasons, is nonetheless meant to be read in its entirety at whatever time of year you happen to find yourself. You can return for a "hands-on" look at the seasonal phenomena at a later date, if desired.

I've tackled the material with both the newly curious and the experienced astronomer in mind. And I have addressed the problem of *definitions* in a way that I think will cause discomfort to neither. We will not pause in the midst of our explorations to define terms such as *galaxy* or *zenith.* Instead, a survey of fundamental astronomical concepts is found in the first chapter, which the beginner is advised to read before embarking on the following sections.

Many of these subjects are strange enough without cloaking them further in unfamiliar language. Therefore the light-year, a wonderful concept explained in the first chapter, is unapologetically our distance unit instead of the parsec preferred by astrophysicists, and the kilometer suffers a rare setback in favor of the mile.

Now, unburdened by technical jargon, come explore a universe brimming with wonder and beauty.

—Bob Berman
Woodstock, New York

Acknowledgments and Dedication

Portions of some chapters originally appeared in my monthly "Night Watchman" articles in *Discover* magazine between 1990 and 1994. My thanks to the magazine's editor in chief, Paul Hoffman, and associate editor Tim Folger for their help and support.

A special thanks also to Will Schwalbe at William Morrow, whose recommendations were right on the mark. If no split infinitives appear in these pages, it's because his pathological crusade prevented a single example from managing to accidentally sneak in. In truth, I cannot imagine a better editor; the clarity of his mind is awesome. Thanks also to Bruce Giffords, Parry Teasdale, Mikhail Horowitz, Geddy Sveikauskas, Andrea Barrist Stern, and Larry Weinberg.

Countless invaluable suggestions came from my biggest supporter and best friend, a longtime writer in her own right—my mother. As one token of my eternal gratitude to this extraordinary person . . .

This book is dedicated to Paula Dunn.

Contents

Spring

Summer

Autumn

Secrets of the Night Sky

Coming to Terms with the Universe

My favorite bumper sticker is the one that says 186,000 MILES PER SECOND—IT'S NOT JUST A GOOD IDEA; IT'S THE LAW!

Most people get the joke: The figure is the speed of light. However, others find the bumper sticker as meaningless as the majority that crowd today's highways with suggestions like HONK IF YOU LOVE ARTICHOKES.

So it is with the educated public. Knowledge of astronomy varies greatly (and is generally very low)—yet people want to be in on the most amazing discoveries. So however you react to the bumper sticker, and whatever your science background, this chapter's purpose is to make basic cosmic concepts clear enough to allow you to plumb the universe's astonishing secrets. If you're already knowledgeable about the heavens, you might treat this chapter as a quiz.

Everyone, not just the raw beginner, can benefit from a refresher on the cosmos because we are exploring a vaster, stranger universe than we thought existed just twenty years ago. An astronomy text from the 1960's is practically useless today. Photographs of Jupiter or Uranus taken twenty-five years ago seem now like smudgy daguerreotypes. The explosion of recent technological advances in the exploration of the universe has left even sophisticated laypeople in the horse-and-buggy age. Many still think of "Mount Palomar" (whose real name was always "the Hale telescope") as the world's largest telescope, even though it was surpassed in the 1970's. Moreover, quantum improvements in electronic light-amplification have effectively made every telescope "bigger." Spacecraft have rewritten the book on planetary knowledge. And wavelengths of light that the human eye cannot perceive (which constitute most of the energy arriving from the rest of the universe) have been monitored by orbiting spacecraft above our planet's obscuring atmosphere.

The knowledge of the universe gathered in the past quarter century surpasses all the discoveries in the entire century preceding it. And that, in turn, dwarfed all previous human learning about the universe since time began.

Happily for those hungering to digest the newer information, astronomy's basic terms and units have not changed. Even better, a single worldwide currency has developed: The tools for probing the universe are universal. In every country, people use the same distance scales, the same designations for celestial objects, the same units of brightness and speed. Even star names have evolved into a common worldwide nomenclature, a particularly notable achievement since every culture started off with its own labels for the stars and their patterns.

Not that the traditional lore has been abandoned. Astronomers in China, India, and Canada still know many of the star names passed down by their ancestors but have also adopted the wonderfully simple, international scheme originated by Johann Bayer early in the seventeenth century. Under this system we call the brightest star of any constellation **Alpha** (the Greek letter *a*) the second brightest **Beta** (*b*), and so on. Using the first dozen or so letters of the Greek alphabet, one can pretty much fake it when it comes to communicating about the stars. The brightest star of the constellation Centaurus? Alpha Centauri! (*Centaurus* gets changed to *Centauri* because the Latin genitive form is used. But never mind: All this is mentioned only so that star names appear less mysterious.) Gamma Orionis and Beta Scorpii are now suddenly seen as sensible designations that reveal their parent constellations and relative brightness.

A sample of cosmic delicacies: planet, stars, and galaxies as seen from an imaginary moon

α alpha	η eta	ν nu	τ tau
β beta	θ theta	ι iota	υ upsilon
γ gamma	ι iota	ο omicron	φ phi
δ delta	κ kappa	π pi	χ chi
ε epsilon	λ lambda	ρ rho	ψ psi
ζ zeta	μ mu	σ sigma	ω omega

The Greek alphabet

There are some exceptions where the brightest star somehow wound up as Beta and a lesser light got promoted to the Alpha spot, as if it had friends in high places—but such quirks hardly matter. And for the sky's two dozen brightest stars, we'll generally use their popular names, the ones we inherited from antiquity. There's no need to call Polaris "Alpha Ursae Minoris"—old friends do not demand such formality.

Every star belongs to one of eighty-eight constellations. **Constellation** is defined not simply as a pattern of stars forming a figure like Orion the Hunter, but as a precisely delineated region of sky. Every piece of the nightly heavens, whether starry or dark, is part of some constellation. The boundaries of this colossal, concave jigsaw puzzle were internationally defined in 1930.

We commonly say Mars is "in" Leo, or a particular galaxy lies "in" the constellation Virgo, but this simply means the object of interest lies in that direction. A planet is typically a million times nearer, and a galaxy a million times more distant, than the stars that seem to surround it. Even the con-

The way we map constellations has evolved with our cosmology. As we stopped seeing Earth and even the Galaxy as the center of the universe, our maps gave up their personalities. The old pictures (at right) *were intended to help us find stars by relating them to myth or experience. Later the pictures became abstractions* (facing page, left), *and finally a grid system enabled us to divide the sky into areas that retained the names of the constellations* (facing page, right).

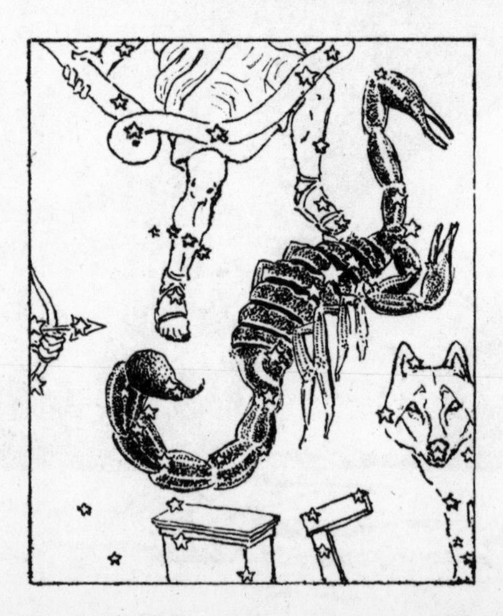

stellation's own stars stand at disparate distances from us. The stars forming the figure of Orion lie as near as 20 light-years from us, and as far as 2,000. You could never actually travel to Orion: The pattern would dissolve as you approached.

We'll use the words **sun** and **star** interchangeably. It's awesome that our sun is merely the closest of the trillion or so stars of our galaxy. Conversely, stars are suns: The illustration on page 164 showing a piece of the Milky Way as myriad stars sprinkled like confectioners' sugar is an image of endless suns. To perceive each star as a sun conveys a grand sense of the generous energy hidden in the hallways of the night.

When we allude to our own sun, we'll say **the** sun, or **our** sun. Our sun is a nearly million-mile-wide ball of nuclear flame, whose powerful gravity holds nine planets in orbit around it. (Actually there are only eight, since Pluto is not now thought to be a real planet—but we'll get to that in a later chapter.)

Most planets are encircled by satellites, and we'll usually follow the popular practice of referring to these as "moons" even though that's officially just the name of Earth's single natural satellite.

Planets revolve around the sun, but a quick way to distinguish between planets and stars might be: If you could touch it, it's a planet. A star is a gaseous globe shining now or in its past by nuclear energy, as if hydrogen bombs were going off nonstop. Thankfully this is not the case with planets, to use Earth as an example. Planets are visible due to the same simple

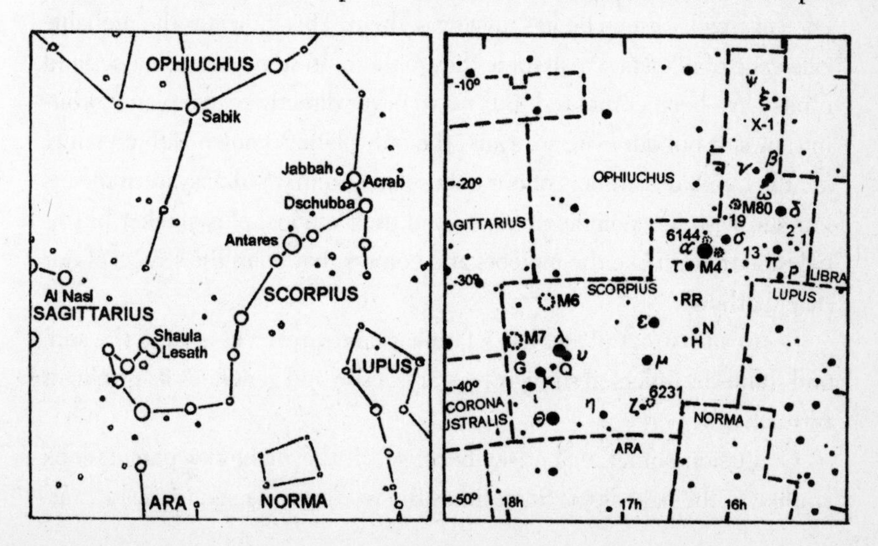

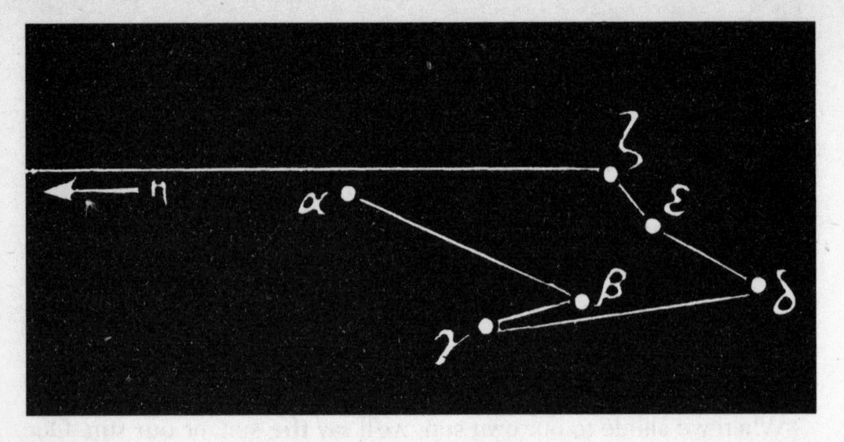

The Dipper viewed from a planet orbiting Gamma Boötis, a star just east of the Dipper and about as distant as we are. Eta (η) is more than 90 degrees away from Delta (δ) and would be part of a different constellation to people on this planet.

process that lets us see a friend down the block: They're standing in sunlight. If the sun should go dark, so would all the planets.

You can already recite the names of every known planet in the universe. Mercury, Venus, Mars, and the rest (Earth, Jupiter, Saturn, Uranus, Neptune, and perhaps Pluto) are the whole show. You'll never encounter any others. Distant stars may be encircled by planets; perhaps they all are. We cannot detect planetlike objects sitting close to the dazzle of their parent stars with present technology. Some indirect evidence, such as the wobble some stars display as they creep through space, suggests the presence of small, unseen bodies tugging at them. This indicates the probable existence of planets. We assume they're there. In some cases, masses and orbits have been computed. But we've never directly seen a planet orbiting any star but our own, so, again, the only planets known with certainty are the familiar members of our own solar system. (**Solar system** means our sun with its encircling planets and their moons, plus smaller bits of celestial flotsam like the meteors and comets that roam the streets of our neighborhood.)

So, to sum up: A planet has a familiar name, revolves around the sun, and shines by reflected sunlight. A star is a sun and generates its own light by nuclear energy.

Confusion sometimes arises because, in the night sky, planets look starlike to the naked eye. Sometimes the word *star* is used to mean "starlike in appearance." Venus, for example, is often called the Evening Star.

In the yawning depths of space, stars arrange themselves in **galaxies**—cities of suns that are good candidates for the grandest things in the universe. Galaxies may assume a spherical, irregular, or lovely spiral design. Each galaxy has at least a billion stars; the major galaxies have closer to a trillion or more. Our own galaxy, a spiral, is called the **Milky Way,** and nobody denies this is the very silliest of all galactic names. That's one reason our home galaxy is often simply called the Galaxy (with a capital G).

Every star you can see in the nightly heavens is merely part of our own Milky Way Galaxy. Beautiful photographs of nebulas (clouds of gas and dust), clusters of stars, exploded star fragments (supernovas)—you name it—depict residents of the Galaxy. You don't have to ask: If it's not a photo of another galaxy as a whole, then the image almost certainly represents an object inside our own Milky Way.

Astronomy, and perhaps cartography, are the only sciences in which distance is always important. Chemists and biologists do not particularly care about the precise distance from their eyes to a chemical reaction or tissue sample; in astronomy, however, we cannot touch, smell, taste, or hear anything beyond Earth and the rocks brought back from the moon. All celestial knowledge arrives in the form of light, and we need to know how far it has journeyed before arriving at our eyes or instruments.

For smaller dimensions like the moon's distance or planetary diameters, we use miles. Actually, astronomers use kilometers, and if you, too, think in kilometers, add about 60 percent to the distance in miles and you'll be close. If you're comfortable with miles, sit back and relax; you won't have to do a thing.

Believe it or not, astronomers like to avoid big numbers, so within the solar system they use the easiest imaginable way of expressing distance: the **astronomical unit.** One AU, as it's called in the trade, is simply the average distance from Earth to the sun, about 93 million miles. This way, when we say that Jupiter is 5 AU from the sun, we can picture its orbit instantly—as five times the diameter of our own. It would be more cumbersome to express Jupiter's distance as 500 million miles. Who could visualize that?

Outside the solar system, we use **light-years.** This is a splendid measure of distance for two reasons. It is large, so we can express the remoteness of stars in reasonable units (a typical star seen with the naked eye is 100 light-years away). It also has the conceptually crisp advantage of using light, whose velocity in the vacuum of space is always constant.

Light's unvarying speed is the fastest thing in the universe. It travels at 186,282.4 miles per second, so if you could run your errands at the speed of light, you could make thirty round trips between New York and California in a single second.

Traveling this fast not for a mere second, but for an entire year, would carry you 5 trillion 880 billion miles: 1 light-year. Despite the inclusion of the word *year,* the term *light-year* is a measure not of time, but of distance, like a mile. It is the distance light travels in a year.

Astronomers often use additional units like the **parsec,** which is equal to about 3¼ light-years, but we're happily going to ignore the parsec along with a few other nonessential measurement concepts.

Most other terms are the same as those encountered in everyday life. Temperature is expressed in degrees Celsius or Fahrenheit, time in seconds and years, angles in degrees. Any circle has 360 degrees, so the inverted bowl of the sky is, conveniently enough, an ordinary hemisphere, 180 degrees. Logically, then, the distance from the horizon to the over-

To put the various distance measures in perspective, we offer this sliding scale. On each line the sun is at the left end and the key distance from the line above is projected downward by the diagonal line.

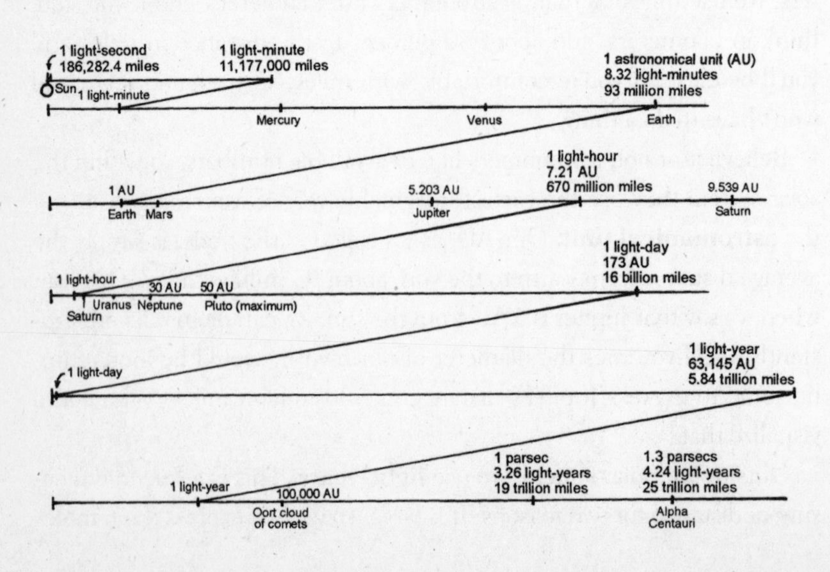

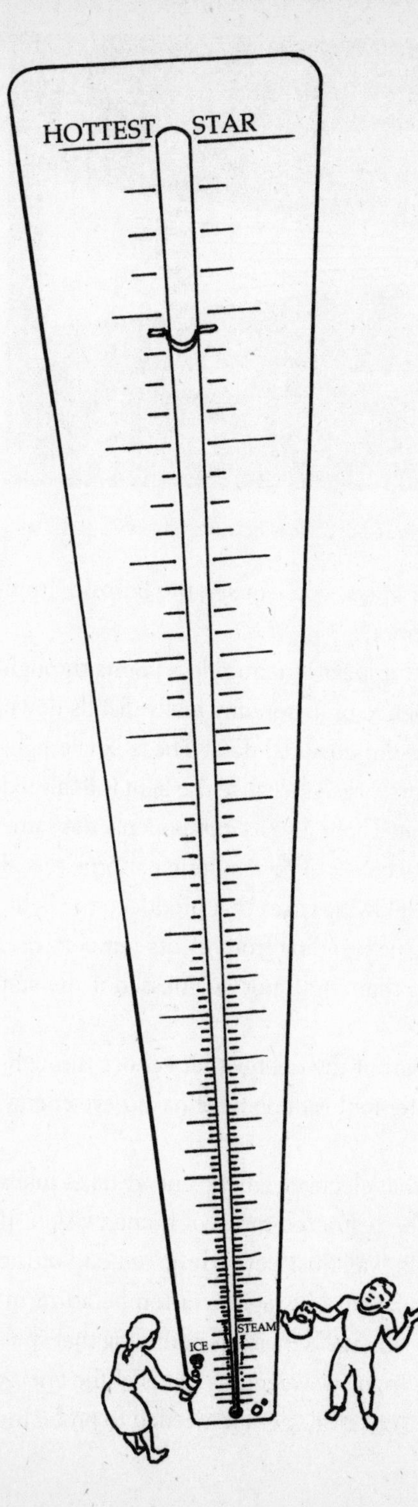

head point, the **zenith,** is 90 degrees. When we care to measure things in the sky, such as the height of the North Star above the horizon, we'll take advantage of this *handy* fact: A clenched fist held at arm's length blocks off very nearly 10 degrees of the sky. (This works for nearly everyone, because if you have a larger fist, you probably have a longer arm as well—pushing that fist far enough from your eyes so that it still subtends 10 degrees of sky.) Try it now, but not in public: Carefully count off nine fists upward from the horizon and see if it takes you to straight overhead, 90 degrees up. If not, simple corrective surgery can make this work. (Or else compress the thumb a bit.)

The **horizon,** in astronomy, is not where the mountains or treetops meet the sky. Rather, it is an imaginary horizontal line lying at eye level in all directions from you, 90 degrees from the zenith. Notice that *horizontal* even contains

The narrow, comfortable range of earthly temperatures gives little hint of the extremes routinely encountered throughout the universe.

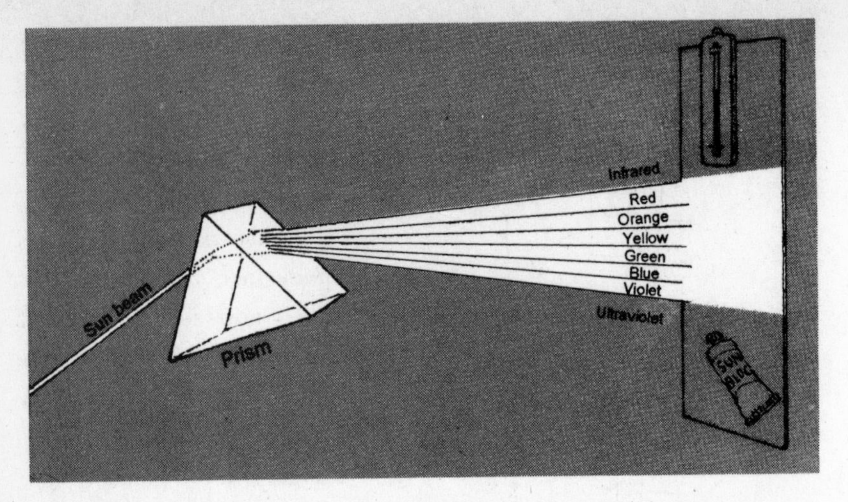

Infrared and ultraviolet are invisible to our eyes, but their effects are familiar.

the word *horizon.* In an airplane or at sea you can see the horizon; from your home you almost certainly cannot.

Because stars are so distant, they appear as featureless points through even the largest telescopes; the science of astronomy really distills down to extracting information from one-dimensional dots! These single light beams tell us volumes about the stars largely because the light is analyzed using a spectrograph, which segregates light into its components the same way a prism breaks sunlight into its colors. The resulting patterns reveal much to us: the composition of the glowing gases that produced the light, whether the source is approaching or receding from us, its temperature, plus a wealth of other information that could not be guessed if the star was merely inspected visually.

Our eyes detect just a narrow part of the energy that echoes throughout the universe, and we usually refer to this rainbow of naked-eye energy, with its familiar colors, as "light."

We all know from grade school that electromagnetic energy takes many forms, most of which lie beyond the restricted range of human vision. If you hold a thermometer on an unlit wall, just beyond the red end of the sun's spectrum projected by a prism, it will register a temperature increase, showing that infrared (heat) energy is invisibly pounding that spot.

Because our atmosphere blocks many wavelengths (parts of the energy spectrum), high-altitude or satellite observatories are needed to probe the infrared and the other invisible light of the universe. When it comes to simple sunlight, however, our eyes are excellent receptors. There's not a lot we're missing, because we see in just those wavelengths that the sun emits most strongly. Our eyes evolved through the ages to see by daylight, and except for the dangerous antilife ultraviolet, largely blocked by our protective atmosphere, we have been given good seats for most of the show.

When we refer to parts of the spectrum detectable only by instruments—X rays, infrared, microwaves, and the like—we'll usually use the word **energy** rather than **light.** Remember, however, that these hidden parts of the energy spectrum differ from visible light only in wavelength: They vibrate at too high or too low a rate to stimulate the receptors in our eyes.

More basic than this, and even easier to understand, is the simple system used to express the **brightness** of things. Modern astronomy imported the scale from the ancient Greeks, who called a bright star "first magnitude" and a dim star "sixth magnitude." Applying modern measuring devices like photometers, we still, as in the days of Plato, consider the very faintest stars seen on a moonless night as sixth magnitude. But the limits have changed: Telescopes and electronic amplification allow us to detect stars that are much dimmer and thus have higher numerals. Even a small telescope can reveal eleventh-magnitude stars. Heading in the other direction, a handful of stars and planets are so bright we assign them zero magnitude, or even minus numbers.

The following table summarizes everything you'll ever need to know concerning the brightness of celestial objects.

The Magnitude System

MAGNITUDE	EXAMPLE	COMMENTS
–12	full moon	
–5	Venus	Brightest starlike object. Can cast shadows.
–1.5	Sirius	Brightest star.

The Magnitude System (Continued)

MAGNITUDE	EXAMPLE	COMMENTS
0	Vega	Only 8 stars are this bright or brighter.
1	Antares	Twenty stars are this bright.
2	Polaris	Medium bright: 60 stars.
3		Dull medium: 150 stars.
6		Borderline visible: 6,000 stars.
9		Binocular limit: 50,000 stars.
12		Barely visible through small telescope; 60,000 times fainter than Vega.
20		Visual limit of Hale telescope; 100 million times fainter than Vega.
25		Faintest Hale telescope stars using electronic amplification.
29		Detection limit of Hubble Space Telescope; 250 billion times fainter than Vega.

The system is set up so that a five-magnitude change corresponds to a hundredfold shift in brightness. But you needn't remember this. What should be clear is that the magnitude measurement addresses only the apparent brightness of objects. A dim star might really be a very luminous one lying far away. The scale merely displays the way celestial objects appear to our eyes.

Speaking of "our eyes," the term **our** will refer to readers living in Earth's north temperate zone. By a happy quirk of population and geography, most of us reside between latitudes 20 north and 60 north, which encompasses the United States, Canada, Europe, and most of Asia. Such commonality of latitude presents us with a precious opportunity that cannot be ignored:

The sky changes only when viewed from a different latitude, or distance from the equator. Its wonders are identical at various *longitudes*, which vary as you go east or west—a geographical spread that happens to embrace much of the developed world.

Move from New York to Tokyo and the sky doesn't change. But go to a Caribbean island and it's very different.

We'll take advantage of our communal latitude by describing common experiences—for example, "a dazzling blue star halfway up the southern sky at midnight"—or saying, "The moon rises upward and to the right." Most readers will easily locate objects under discussion because the orientation of the sky is identical for the vast majority.

But be aware such descriptions will not be accurate if you're near the equator or in a Southern Hemisphere country like Australia. All objects in the southern sky will then be much higher up than stated, while northern targets will be lower, or perhaps even invisible. The motions of the sun and stars will be different as well.

For many of our topics—twilight, blackness, other dimensions, and cosmic birth, to name a few—the location of the reader is irrelevant. In fact, you can fully enjoy almost all of the book without stepping outdoors at all.

On the other hand, after learning about an amazing object, it's fun to see it for yourself. For those inclined toward a more hands-on approach, the subjects, as explained in the Preface, are arranged in the order in which they present themselves throughout the year. Of course, the reader is under no constraint to follow them in that sequence or, for that matter, in any order whatsoever.

No instruments are needed for probing these extraordinary entities of our universe. However, a few subjects—the moon and Jupiter, for instance—are dramatically enhanced through the use of binoculars or a small telescope. Guidance in selecting and using these instruments is found in the Appendixes.

One note about style. Rather than using the cumbersome "he or she" when alluding to people, we'll simply alternate pronouns. If an astronaut

A tale of two cities: From New York the sun appears to travel rightward, while its motion is forever to the left for residents of Sydney, Australia.

is called "she," the next one will be "he." Aliens are "they."

And now we are ready. Armed with commonality of latitude and language (if not bumper stickers) and sharing a love for the curiosities of the cosmos, we here begin an exploration of some of the most amazing things in the universe.

Winter

The famous moon illusion (opposite) *makes the moon seem enormous when near the horizon. Okay, so we exaggerated a bit.*

It's About Time

The midnight stars, like horses returning to their stables, come back to the same position every New Year's Eve. The mad blue twinkling of Orion and the Dog Star count off with tireless vigilance the years that whoosh through human lifetimes like windblown leaves. Silent, detached, they provide perfect contrast to the cacophonous celebrations roaring below them every December 31.

The philosophically minded might see human merrymaking as a show of bravado against the reality of our mortality. Others see New Year's Eve as a celebration of romance. Still others feel its nostalgia. I, for one, view New Year's Eve as a unique curiosity: the sole occasion when everybody is focused on time, one of modern life's obsessions. And with the stars returned to their starting points, and the year's clocks reset to zero, it certainly is an appropriate moment to explore the amazing dimension of time itself.

But pursuing the subject is a little like chasing the white rabbit with the pocket watch, since *time may not exist.* At least that's what many physicists have concluded, echoing thoughts long debated in the labyrinthine corridors of philosophy. But if time does have reality—or even if we all merely act as if it does—then it is odd indeed that in a society that prizes punctuality, misconceptions about time hatch as readily as cuckoo eggs. *And New Year's crowds shout and hug at the wrong moment.*

As one who keeps his watch precisely set to atomic clock signals, I find it remarkable that televised New Year's Eve celebrations invariably flash twelve midnight when it isn't. Especially since the ability to consult an essentially flawless system, so astonishingly accurate it scarcely could be improved upon, is available to everyone.

Such perfection has obviously not trickled down to the huddled masses yearning to be on time. When one switches on a typical cable weather channel or, worse, calls a local time number, the announcement is nearly certain to be off by anywhere from a few seconds to half a minute or more.

Should that matter? Aesthetically, yes, since a watch set to the *right* right time is a device synchronized with the rotation of Earth itself. For decades, the world's clocks have been commanded to ticktock in cadence with the stars. Keeping time to the spin of our planet has added an elegant touch, for the universe thus parades overhead in rhythm with the timepieces strapped to our wrists. It's almost as if we're bound to the clockwork mechanism of the solar system.

In a little less than a day (23 hours, 56 minutes, 4.1 seconds), Earth rotates 360 degrees, so that the same distant star is once more directly overhead. This is a sidereal day, but the sun is still about 1 degree shy of the zenith. In 24 hours (on average) the solar day is complete. (The angles are exaggerated for clarity.)

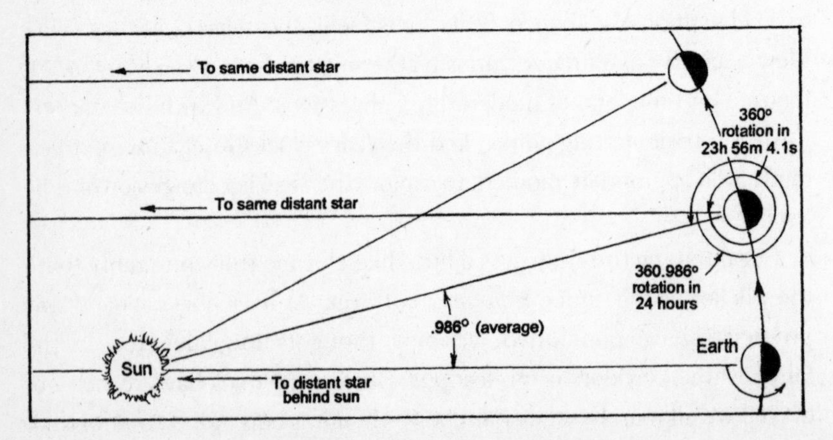

It wasn't always this way. Any regularly recurring phenomenon can serve as clock or calendar, and through the ages the diurnal sequence of day and night, the phases of the moon, and the returning of the seasons have been obvious choices—just as in today's society, dates like April 15 produce dependably cyclical reactions.

From the beginning, though, we had to make a choice. We could devise a timepiece that marches in step with our planet's rotation of 23 hours, 56 minutes, and 4.1 seconds, which would allow the stars to rise and set at the same moment every night. Or we could instead make clocks keep pace with the average position of the sun. This solar day amounts to Earth's rotation plus an extra four minutes, because when Earth has completed one spin, the sun's in a slightly different direction since we've also been orbiting it.

It was an easy decision. We chose the sun. We devised clocks whose hands perform a single spin as the sun circles our sky once, and to heck with the stars, which then must rise and set four minutes earlier each night. The stars, by our own decree, thus do not stay in tune with our clocks, and nobody's lost any sleep over the matter.

In the 1950's it was decided that Earth's spin should be the standard after all, rather than vibrating quartz crystals or any other timekeeping method. Of course we weren't about to toss out all the world's clocks and watches; we'd still keep solar time. But we'd periodically adjust all timepieces to keep the planet's spin firmly locked in step with our chronometers. If our planet sped up or slowed down for reasons of its own, we'd regulate billions of clocks accordingly. We'd add or subtract "leap seconds" on the final day of June or December, and Heaven and Earth would swim in temporal harmony. In between, the passage of time would be monitored by atomic clocks.

The atomic clock! A wonderful device that remains accurate to one second per thirty centuries, it uses the vibrations of subatomic particles within the cesium atom to maintain perfect regularity. Every second it makes 9,192,631,770 oscillations (a number I typed without needing to consult a reference, for it's engraved in the brains of all time fanatics such as, alas, myself).

This love of precision need not extend to other areas of life. More often than not my desk is buried in clutter and I am not at all bothered by the black snake that lives beneath the clothes dryer. In my rural area, nature's vagaries unfold with reckless imprecision and it's pleasant enough to be swept away by them. Time is something else.

When I recently went to the town library and found the door locked even though my watch read four minutes to closing, it illustrated the prerogatives of precise timekeeping. I rattled the door until the librarian appeared, who pointed with finality to the clock on the inside wall. It read 5:01.

Responding to further rattling, she unlocked the door just long enough to say, "It's after five"—and here is where knowing the right time comes to the rescue.

"It's four minutes *before* five," I insisted, tapping at my watch.

"How can you be so sure?"

"Because I set it twice weekly to shortwave signals from the cesium atomic clock at the Naval Observatory. It might now be off by two seconds, but no more than that. If you like, we can call them on your phone. . . ."

She let me in—probably because it's generally wiser not to quarrel with a fanatic. The point, however, is that few areas of contemporary life are cut and dried. There are always different sides to every question, shades of gray, relative truths. Not so with time. Here there is no ambiguity at all. There is only one right time. Anything else is wrong. There is no discussion. There is nothing else like it; it's intoxicatingly unambiguous.

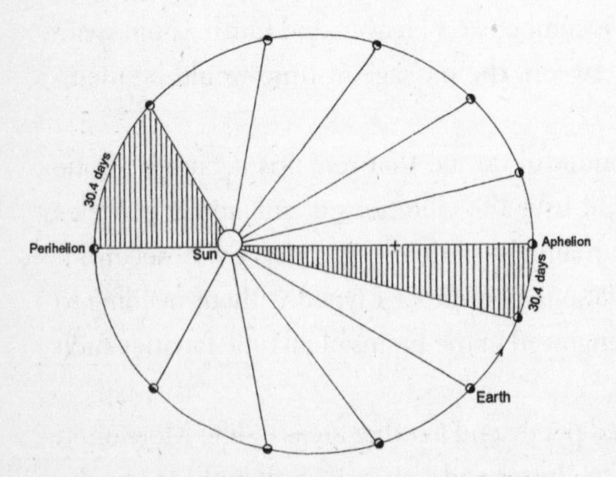

According to Kepler's second law, Earth's orbital radius sweeps over equal areas during equal times, so we travel faster and cover more celestial real estate in December and January, when we're nearest the sun. One consequence is that Earth must rotate a bit more each day at that season before the sun is again overhead (noon to noon), making the day longer.

All this is undebatable (and even poetic), but it's not without misconceptions that seem to reach a climax at New Year's Eve. For example, that leap second inserted at the end of most years' final minute is usually overlooked at New Year's celebrations—meaning that even without further timekeeping sloppiness, all that kissing happens a second too soon. But the error goes beyond a Times Square overflowing with a million premature hugs and shouts.

Most people understand—when they stop to think about it, which is never—that leap seconds are needed because of our planet's irregular spin. But they wrongly assume they're also necessary because Earth is slowing down. This is a widely held misconception, routinely repeated in print. But if we were slowing that rapidly we'd have come to a frozen halt long ago. Earth, thanks to the moon's tidal pull on the oceans, which in turn act as a giant brake on the world's beaches, is slowing by merely one five-hundredth of a second per century; certainly no reason to meddle annually with clocks.

No, the fault is not in the stars. The necessity of adding so many leap seconds originated during the 2¾-year period in the late fifties when Earth's rotation was carefully monitored to find the exact length of the

The combined effects of Earth's tilted axis and eccentric orbit cause the length of the solar day to vary by a total of fifty seconds over the course of a year. P and A represent, respectively, the times of year when we're closest to and farthest from the sun.

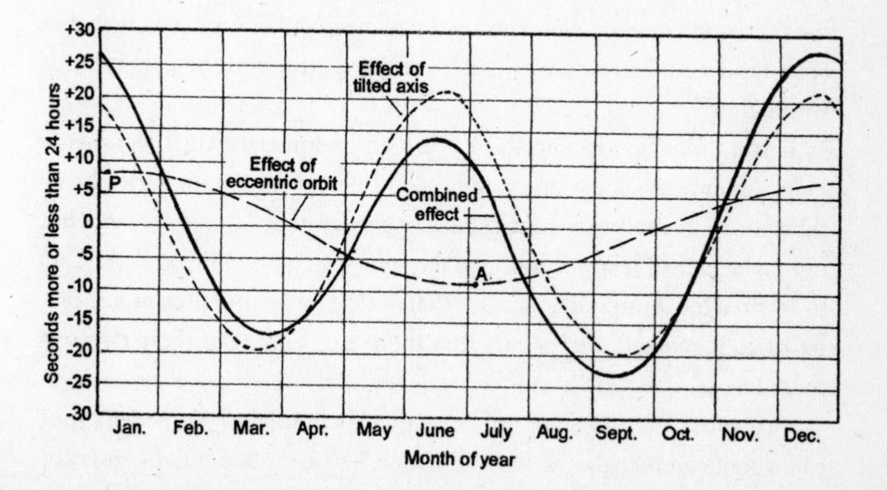

day, a prerequisite for establishing the current system. Either Earth spun anomalously slowly those few years, or else the measurements were not precise enough: Ever since, all our clocks have been running slow! That's why a total of seventeen seconds have been added as of 1994—but none ever removed.

The first leap seconds were introduced in 1972 when two were inserted, making that the longest year in history. (And it probably felt that way to presidential candidate George McGovern.)

None of this would have mattered just a century ago, for no timepieces were precise enough to cast slander on the planet's spin. A watch accurate to a second per day would have made any physicist whirl in ecstasy. Today's cheapest quartz watches double that precision, and they'd do even better if quartz didn't vibrate slightly differently with changes of temperature and such.

It's astonishing that the quartz watch even works at all. It has long been known that if you start a little chunk of quartz oscillating it generates a tiny electric current—the piezoelectric effect. Who was the genius who discovered it works the other way around as well: If you apply a tiny voltage, the mineral itself vibrates at a constant rate without ceasing? The constancy is what does the trick, the secret behind the quartz timepieces first used in 1928 and now attached to billions of human bodies as if the mineral were somehow essential to health.

Quartz watches suddenly made the old spring-driven mechanical timepieces seem like hourglasses. Quartz, constructed of silicon and oxygen, the two most common elements on the planet, could now keep track of its spin. How could anyone wear such a clever device and then set it to the wrong time?

If you do decide to join the world's chronomaniacs you'll get lots of help. Already available are watches and clocks that automatically set themselves to shortwave time signals engendered by atomic clocks. They even take care of leap seconds and daylight saving time. If you don't want to spend a few hundred dollars for that automated perfection, or just prefer to do it yourself, feel secure that there are people out there devoting their lives to assist you in your struggle for accuracy.

First you've got an agency in France that keeps track of our planet's spin, called, not surprisingly, the Earth Rotation Service. Then there's America's official timekeeper, the National Institute of Standards and Technology (NIST), formerly called the Bureau of Standards. Its Time Division choreographs an awesome effort toward ever-improving accuracy (although several staff members, including Don Sullivan, its chief, confessed to me that they keep their own watches a few minutes fast to be on time for meetings).

If you can't spare the fifty thousand dollars for an economy-model cesium atomic clock, not to worry. NIST has dozens that transmit the right time to within a hundred billionth of a second. A little sloppiness sneaks in only because it takes time for the signals to reach you.

Everyone who's ever twiddled a shortwave radio has discovered the loud, insanity-provoking time beeps at 5, 10, and 15 megahertz and other frequencies, broadcast on NIST's WWV station and accurate to about one thousandth of a second—a millisecond. Not bad, but real nuts for precision can now turn to the WWVB station at 60 kilohertz, which transmits by ground waves that arrive even faster, doubling the accuracy.

No shortwave? No problem. Special phone lines have been set up to reduce switching delays, allowing you to plug into an atomic clock with a simple phone call. The number (303) 499-7111, during off-peak hours, costs even less than the modest fifty-cent charge for dialing the U.S. Naval Observatory's cesium clock at (900) 410-TIME. This may be the least sexy and most predictable "900" call you can make. And imagine: two separate atomic clocks competing for your time and attention. Either way, you get the time to within one thirtieth of a second.

Got a PC with a modem? Then you can receive a wonderful automated computer time service known acronymically as ACTS. If you're not very computer-adept, NIST will send you the software to run ACTS for thirty-seven dollars; call (301) 975-6776 and order research material 8101. Some shareware programs do this as well, simply and beautifully.

Then you're in business, for ACTS does more than merely let your computer grab the right time for itself. It instructs it to echo back the signals for a few seconds while ACTS automatically calculates the round-trip delay, divides it by 2, and then offsets its own time signals, so that your computer now locks onto the right time to the nearest millisecond! Reached by phoning (303) 494-4774, the service is free for the call, so there's no excuse anymore for an inaccurate New Year's celebration.

Instead of going to a party where your new year begins on a sloppy note, just gather your friends around the computer screen instead of the television screen, raise your glass, and welcome in the future.

Or, if you can spare the time, just sneak away from the shouting crowds and listen to the mute ticking of the night's stars, the grand clockwork behind it all.

Baa, Baa, Betelgeuse

Quick—name a star.

People's minds work alike when asked to recall stars, for only a handful of stars are famous enough to compete with the Hollywood variety for space in our memory banks. If you can name three, they'd probably be Polaris (the North Star), Sirius (the Dog Star), and Betelgeuse.

Betelgeuse sticks in children's minds as soon as they hear it, and it never leaves. Of course, the hit movie about a mischievous poltergeist named Beetlejuice only strengthened its place in the public consciousness. But this strangely named, distant sun of the constellation Orion is more than a phonetic oddity; it's one of the most incredible objects in the known universe.

Clearly, however, we can't proceed without a note about pronunciation, since Betelgeuse is the foremost gremlin of astral fluency. Most people say it the way the movie did, as if it were the result of a careless summer footstep. While that is not strictly incorrect, the preferred pronunciation is BET'l-juice. Any further options, like bet'l-GOOZ, are just simply wrong, not to mention increasingly bizarre.

From where could such an odd name have come? One is tempted to fantasize about a noble pet bearing that title, a beloved steed some an-

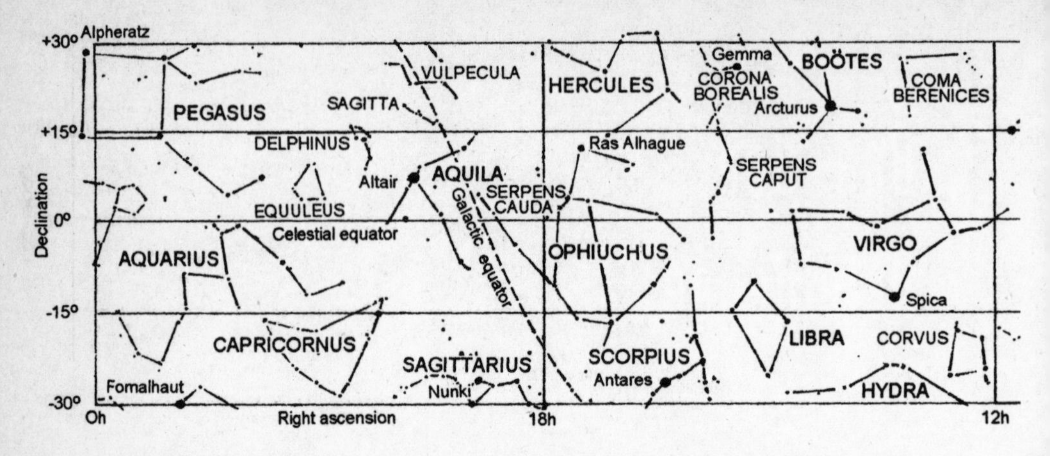

The celestial tropics: The stars near the celestial equator are visible from almost every spot on the planet. The brightest were chosen for the navigation tables familiar to sailors and pilots.

cient rider lost in battle and immortalized in the heavens. Or perhaps it was the corrupted name of a lover whose sad tale was retold for centuries in the Arabian desert. Alas, the truth is sillier than fiction.

To the ancient Sumerians, the star pattern of Orion did not represent a lionhearted hunter at all, but its very opposite—a sheep. And Betelgeuse, it's embarrassing to tell, literally means "the armpit of the sheep." It might be the least appealing name in the universe.

But Betelgeuse's lowly ovine genealogy is scarcely more relevant than a dry scientific profile of its characteristics; the star is so awesome it lives in its own isolation at the fringes of human comprehension.

Simply put, Betelgeuse is the largest single thing most of us will ever see. Yes, a galaxy is larger, but that is a collection of stars. Moreover, not a single galaxy is bright enough to appear in the light-polluted skies over much of the world. Betelgeuse, on the other hand, is brilliant enough to bulldoze its way through the milkiest urban conditions. And because it hovers almost directly over Earth's equator, it is one of the few bright stars that can be seen throughout the world. It's hidden only from the questionably sane community of scientists who willingly reside at the South Pole.

Why should Betelgeuse and a few other stars be seen from everywhere? The way stars' appearance varies with the observer's location is re-

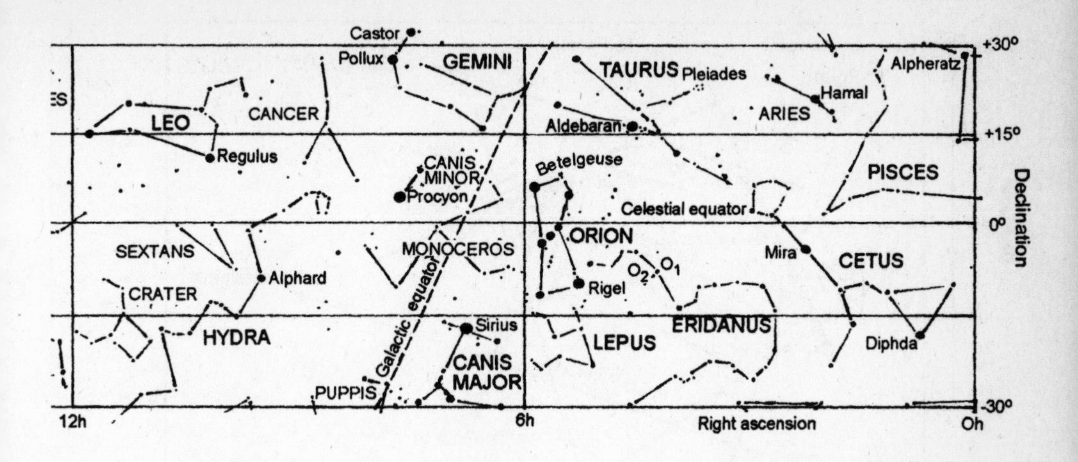

ally very easy to visualize. If you spin a globe after inserting thumbtacks at various positions, you'll quickly perceive that the visibility of the surrounding "universe" depends on your earthly site. From the equator, no part of the external universe is permanently hidden since you see everything around you over the course of a complete spin. But perch yourself near the North or South Pole and the waistline of Earth forever blocks half the cosmos.

The converse is also true: A star that happens to be situated over the equator is seen by everyone. A star over one of the poles—the North Star, say—is forever cloaked from people of the opposite hemisphere, obstructed by Earth itself.

From the latitude of Europe and the mainland United States, about a fourth of the cosmos never rises. Major luminaries concealed from our view include the nearest stars (the Alpha Centauri triplet), the night's second brightest (Canopus), and the Southern Cross, that smallest but brightest of all constellations.

Equatorial constellations are the Esperanto of space, humankind's cosmic common denominators. Orion, straddling the equator like a diplomat, is seen from everywhere. Only one bright star—Procyon—is a whisker closer to dead center than Betelgeuse (and by chance it's a fellow pilgrim of the winter sky).

The immensity of Betelgeuse can be grasped by a simple scale model. If we picture this leviathan as a globe big enough to enclose a twenty-story

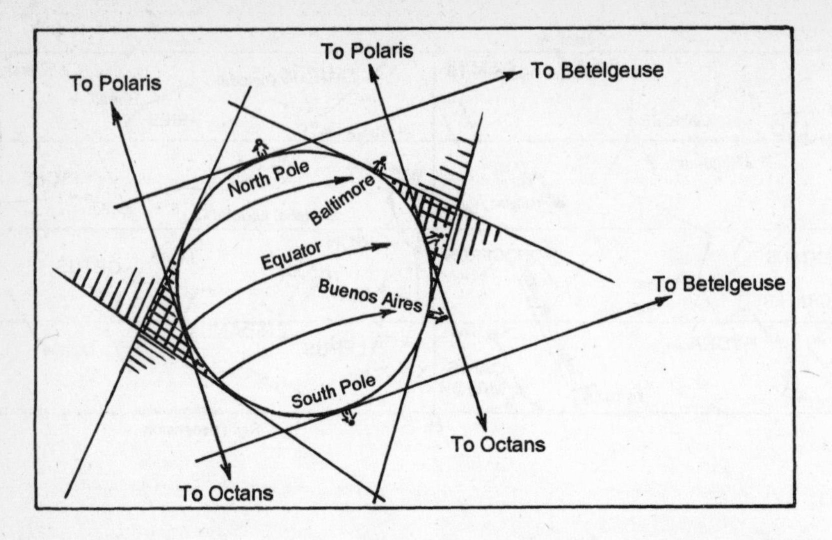

The bulge of Earth's equator hides southern stars from Baltimore's view and northern stars from Buenos Aires', but everyone on the planet can see equatorial stars like Betelgeuse. (Octans is the constellation nearest the south celestial pole.)

building—take a moment to imagine it—then on the same scale our planet, Earth, would be the period at the end of this sentence.

If that star were an empty jar and we could unscrew its lid to pour in balls the size of our planet at the rate of a hundred a second, we couldn't fill Betelgeuse in thirty thousand years.

You'd think something so colossal would be sluggish, since we perceive obese people as less nimble than, say, ballerinas, and sports cars as more agile than trucks. It strains our imaginations to picture this bloated beachball from another dimension doing anything but floating inertly in the pool. How can something so large exhibit zippy animation? But it does: Betelgeuse breathes in and out, changing its size wildly.

Usually it's just plain enormous, some 200 million times bigger than the volume of the sun. Then a burst of energy at its core drives its surface outward until it becomes a billion miles wide, huge enough to occupy our solar system out past the orbit of Jupiter. But whoops, it went too far. Its central furnace cannot sustain such a size. (Who could? Ever try feeding a teenager?) The force of gravity forces its surface to collapse, which continues until the process of shrinking goes too far and creates too much

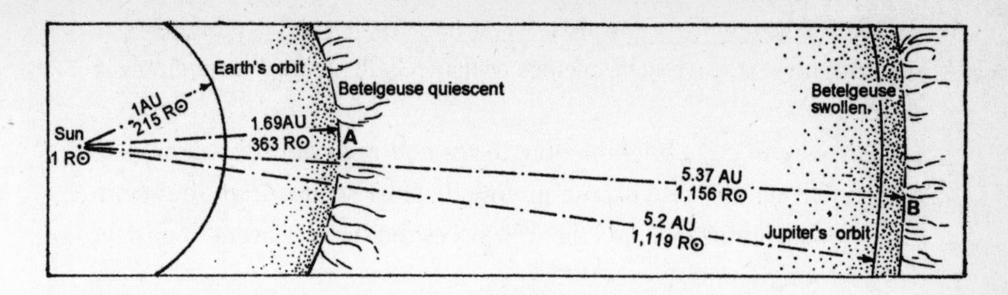

Betelgeuse's radius fluctuates from A to B and back. The radius of the sun (1 R⊙) registers as a mere point on this scale.

heat, so that it rebounds like a trampoline. Out it goes again, to perform the trick all over.

Many stars lead such unstable lives, but red giants like Betelgeuse have a particular propensity for this yo-yo existence. There is no known red supergiant that does not grow and shrink like Alice, varying its brightness all the while.

Betelgeuse not only flickers continuously like a campfire in a storm, but also displays an irregular fourteen-month period when its brightness

The seventeen brightest stars, with the constellation where each is found. Like a bad boy, Betelgeuse keeps cutting in line.

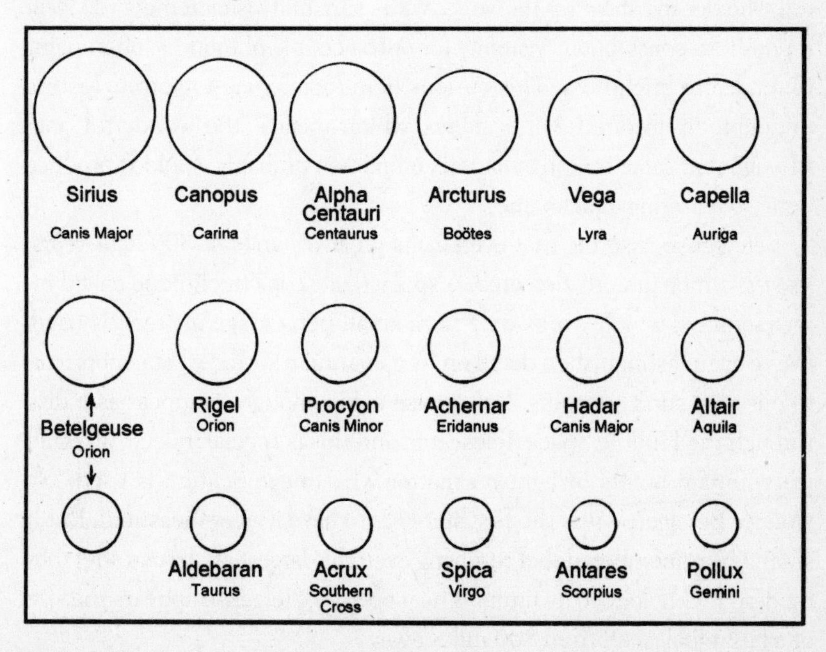

rises and falls greatly, a cosmic tide. At its crest it can become heaven's sixth-brightest star, while its average brilliance still places it respectably in eleventh place.

Here we can call a brief time-out to see how red giant stars allow us to be celestial sleuths, and expose profound stellar secrets from the seemingly random sprinkle of dots that composes the starry heavens. Consider this line of reasoning:

Ordinary red stars are so faint that we can't see them at all, not one example, with the naked eye.

Therefore if we do notice a reddish star—and there are dozens whose ruddy tint is obvious to the naked eye—we can conclude that it's a giant. Only a star with a huge radiating surface can compensate for its dimness per square foot.

Now that we've deduced that the star we've spotted must be a red giant, we know it varies, and the way it appears tonight will be different from the brightness it shows on other occasions. We've unearthed a chameleon. All this—just from noticing the star's color.

The red giant Mira, for example, floating in the southwest during early winter nights, is usually not there at all! Mira's position is marked on maps, but normally it's just a blank spot of sky, like the political hack on the payroll who doesn't show up for work. Mira's variability is extreme; every year it pops into conspicuous visibility for only a couple of months, often rising to moderate brightness. This curious here-today, gone-tomorrow feature undoubtedly inspired Mira's name, which means "the wonderful one" (though that same trait in a marital companion probably wouldn't produce quite so admiring a nickname).

Betelgeuse is so big that even at its great distance of 500 light-years, its size can be directly detected. A special telescopic technique called interferometry, which works only for a small percentage of stars, is used. We've even distinguished detail on its gargantuan surface—starspots akin to our own sun's sunspots. Betelgeuse is big enough to appear as a disk through the Hubble Space Telescope, and this is special indeed; all other stars remain points of light no matter what magnification is used. No wonder Betelgeuse was the first star ever to have its size measured. But it should be remembered that resolving even this largest of all stars stretches modern technology to its limits: The globe of Betelgeuse appears the size of a basketball seen from 800 miles away.

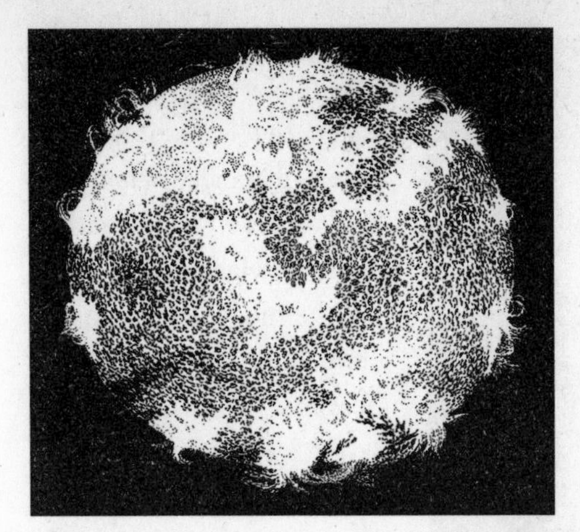

Betelgeuse's huge surface is so far from its center of gravity, its midriff bulges like an unsuccessful dieter's.

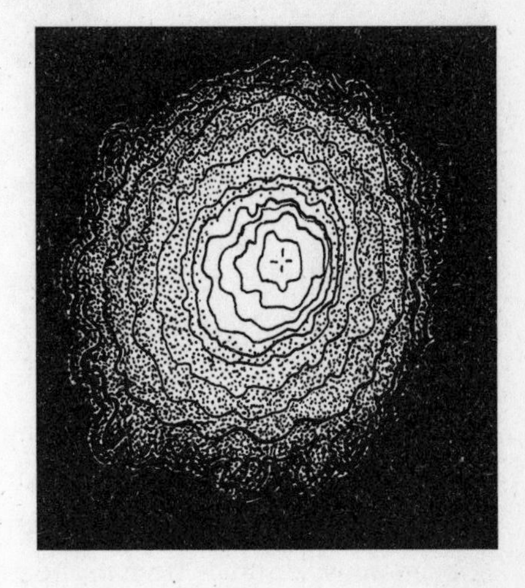

The potassium gas cloud surrounding Betelgeuse (the star is invisible in the center of the cross). The dark lines are contours from a computer image obtained at the Kitt Peak National Observatory.

If your hunger for superlatives has become jaded, your imagination whetted for something even larger than Betelgeuse, consider the tenuous shell of material discovered in 1978 that encases the star like a cocoon. It is made of potassium, like a mineral supplement for some Brobdingnagian hypochondriac. This immense globe extends eleven thousand times farther from Betelgeuse than we ourselves sit from our sun. It is nearly three hundred times the diameter of our entire solar system out to the orbit of Pluto. To construct a scale model of this potassium shell would require it to be a ball three hundred times the height of the Empire State Building, a globe so high that it would extend nearly to outer space and cover the area from New York to Philadelphia. Only then could our Earth be placed alongside it for comparison—once again merely the size of the period ending this sentence.

But big doesn't necessarily mean obvious. If you're a novice and want to locate Betelgeuse, simply find the only bright-orange star above or left of the famous belt woven of the conspicuous three-in-a-row stars in the constellation Orion. And while you're at it, lower your eyes a bit to nearby

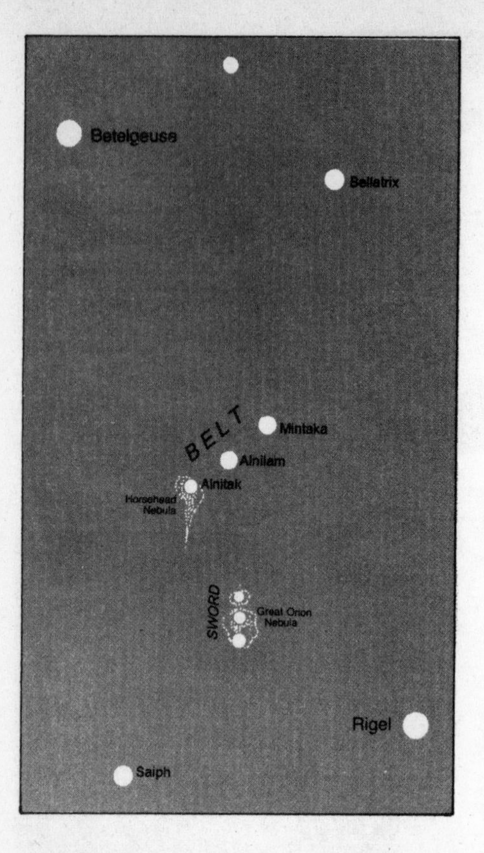

Orion the Hunter

Rigel, the brilliant star an equal distance *below* the belt. These two dazzling stars, Betelgeuse and Rigel, are the most famous pair of contrasting pastels in all the heavens.

We didn't reserve a chapter for it, but Rigel (RYE-j'l) deserves at least honorable mention on a "most amazing" list. This sapphire lighthouse at the foot of Orion serves as the perfect complement to Betelgeuse and, like its ruddy friend, is a supergiant. But with Rigel the astonishing characteristic is brightness rather than size.

Rigel is quite possibly the most luminous star the unaided eye can behold. A blue-white arc welder in winter's icy night, it shines with the same light as fifty-eight thousand suns, an unimaginable intensity. Even at its immense distance from us, nearly twice as far as Betelgeuse, it still outshines all but five of the night's stars. Four of these—Sirius, Alpha Centauri, Arcturus, and Capella—are all nearer than 50 light-years, yet Rigel gives them competition from a perch almost 1,000 light-years farther away, so remote that it sits on the next spiral arm of the Galaxy! If Rigel were as close to us as the others, our nocturnal landscape would tingle with sharp, alien Rigel shadows, and the night sky would always be as bright as when a full moon is out. Most of the universe would disappear from view.

Despite Rigel's supplying a vivid color contrast, it's obvious that, contrary to myth, Betelgeuse is not really a red star at all. Because it is technically termed a red supergiant, some observers expect it to live up to that

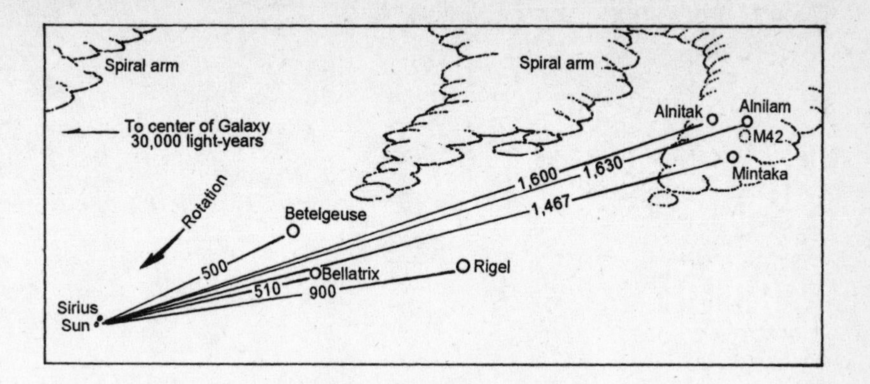

The stars of Orion lie at many different distances (here given in light-years and drawn to scale).

title and be as red as a stop light, or at least a glowing ember. But if you could match Betelgeuse to a color chart at a paint store you'd find it slightly on the orange side of yellow.

While fantasies of painting the largest surface in the known universe might seem an attractive plum to a defense contractor, most of us would vote for leaving Betelgeuse exactly as it is, because it's the only major star of Orion that isn't blue-white. The ancient Hunter is certainly beautiful, but its stars seem tinted from the same can, as if the Eternal Painter made a close-out purchase a few million years ago. Only Betelgeuse supplies variety. In a constellation of young suns, it alone is the stellar senior citizen. Its color is the hue of a giant star in its "golden age" of evolution.

When you are as bright as Rigel or as big as Betelgeuse, you pay for such excesses—or glory—with your life. Our glance at Orion takes our eyes to youth, to a region where stars are still being born, and simultaneously to old age. Supergiants such as Betelgeuse live in a curious netherworld where birth and death occur in the same cosmic breath. Their youth *is* their old age. Betelgeuse, condemned to an early passing, has a further lifetime of less than a quarter-billion years, 5 percent of the time until our own sun's retirement.

Still, that's long enough—more than ample to enjoy it every winter of our lives, and to reveal to our children and grandchildren the topaz flame in Orion's shoulder that is the largest single thing they will ever see.

The Cradle of Night

Sooner or later every sky explorer turns to an immense cloud of gas and dust—the Great Orion Nebula. Its lure transcends its exotic, ghostly fluorescence, a neon excitation from a striking six-sun system sitting in its center. Here is something pristine, powerful, and mysterious as a baby's dreams: It is birth itself.

The Orion Nebula is a cosmic nursery, peppered like mustard seeds with countless infant suns. This womb in space is so immense—30 light-years across—that our fastest rockets would require a half-million years to cross it.

But a strange languor greets anyone whose telescope is pointed its way; the nebula seems frozen and inert. This apparent lethargy stems from our own bias, for its life unfolds on a scale that makes earthly activities seem like the nervous flitting of gnats. Laying dazzling blue eggs like an immense celestial robin, the nebula alters its shape over the span of aeons, as if to hide its intentions from the transient eyes of human generations. But if we could construct a movie of the Orion Nebula by running a single frame every millennium or so, the resulting time-lapse photography would take our breath away.

We'd then see this stupendous swirling ocean as a living mosaic of crimson, emerald, and blue, metamorphosing in greater and lesser ways, containing within it knots of colorful dancing eddies that magically sprout the fires of newborn suns.

The heaviest new stars ignite with intense sapphire flame, a color that is the universal signature of stellar youth and, ultimately, premature death. If we could somehow pay the nebula a call, if we could tiptoe into the nursery for a closer peek, we would also uncover a myriad of fainter lightweight stars—the more sedate and commonplace orange suns that are the Smiths and Joneses of the Galaxy.

Simple binoculars allow the Orion Nebula to be glimpsed firsthand. Conveniently, it floats in the very center of the "sword" of Orion, which dangles from the leftmost star of Orion's famous belt. It's an obvious gray-green smudge, an instant contrast to the crisp starpoints of the constellation itself. In frosty winter skies it stands out sharply, an enigmatic blur aptly symbolizing the mystery of cosmic genesis.

It is here, facing Orion's newborn suns, that we might appropriately turn to the question of our own birth.

The Orion Nebula is a massive cloud of gas and dust 30 light-years across. Ultraviolet radiation from young stars hidden in the cloud makes the nebula glow like a neon lamp.

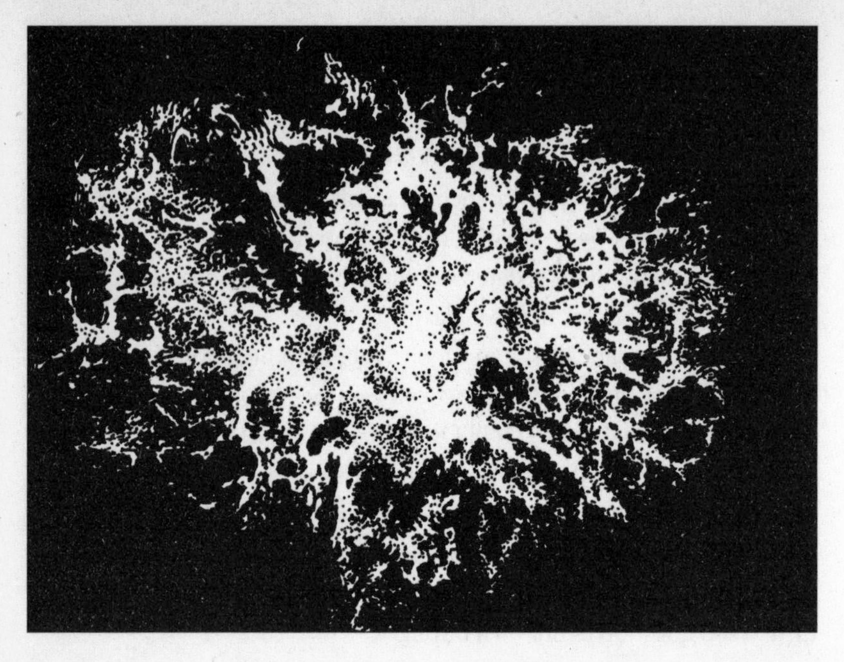

The supernova of A.D. 1054 as it appears today. Our bodies came from a similar explosion.

Most youngsters, at some point, ask where they come from, and the reply may be religious or scientific depending on the inclination of the parent. When we make that same inquiry about our solar system, a stunning scientific explanation, one so lovely that the science itself seems sanctified, is now known with reasonable certainty.

Our ancestral roots, those of every human, animal, mushroom, and bit of moss, trace back to atoms of a single cosmic cloud and a brilliant unknown star, from whose marriage we issued. Like a Medusa with a trillion heads, we lived then as a single entity.

It started, oddly enough, with death. And not just any death. Not a quiet dying that might pass unnoticed in the hallways of the Galaxy. This was the explosive downfall of a massive blue sun, a supernova. The sudden blinding brilliance of its last cosmic moments briefly rivaled the light of a billion suns.

We were that exploding star. Our atoms were its body. We—as that star—had lived brightly but, like blue suns everywhere, had died in childhood. After just 100 million years, a mere fiftieth of Earth's present age,

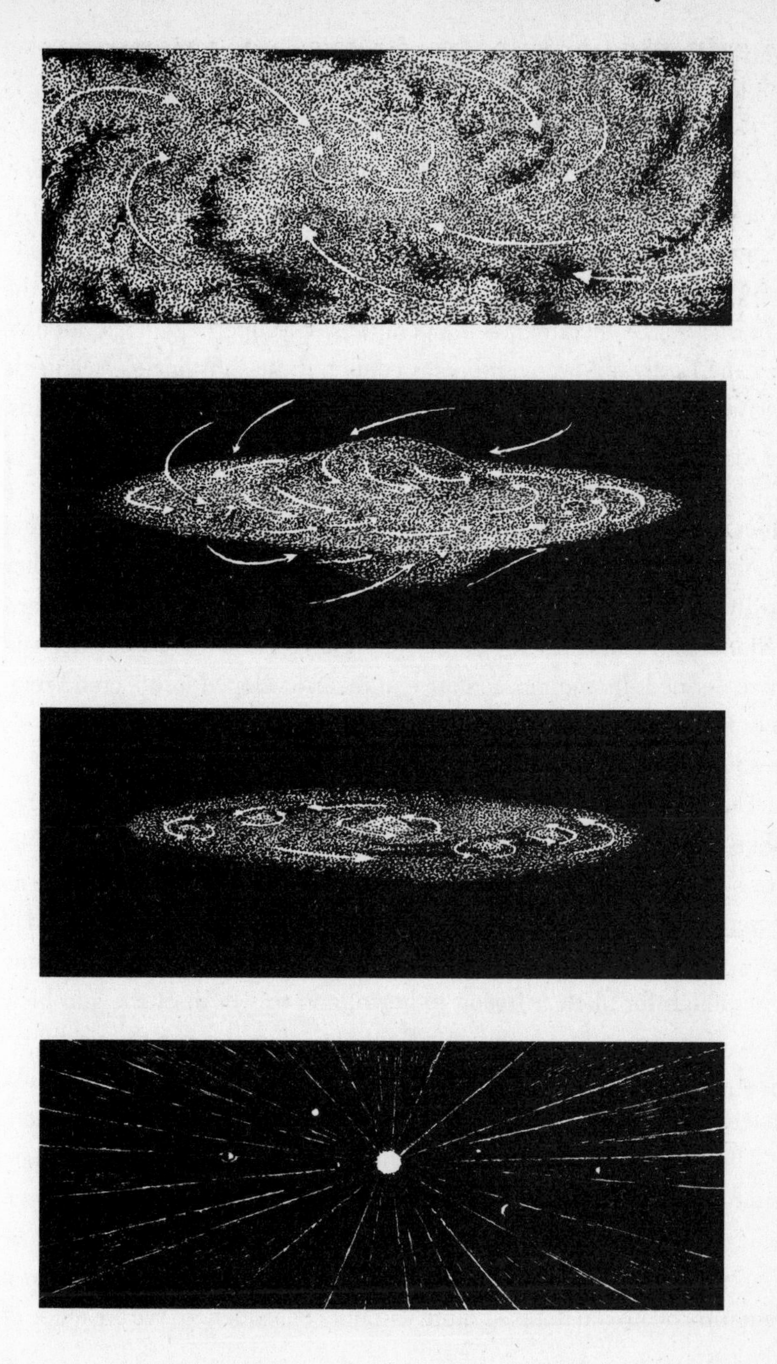

The birth of the solar system: The nebula contracts and flattens. Residual lumps condense into planets.

our star had used too much of its central nuclear fuel. It paid dearly and suddenly for its profligate consumption of hydrogen.

The explosion created such high temperatures that the debris from the crumbling star hurtled in all directions at speeds that exceeded a bullet's by a thousand to one. New atomic elements arose like phoenixes. The ephemeral furnace did what ordinary stellar interiors cannot: forged elements heavier than iron—for such things can be constructed only in the fantastic but temperamental temperatures of a supernova. That our own sun and Earth and even our bodies contain these substances—elements like the iodine in our thyroid glands—is the Rosetta Stone of our origins. It's proof of our family tree; our progenitor was a supernova.

The star's fragmented body flew through space until it contacted a quiet nebula, a gas cloud, where the two performed a mating minuet that merged their masses so that gravitational forces could go to work. They slowly rotated, and contracted, and each new collapse increased the gravitational pull still further. The new combined nebula grew smaller and more defined. Its motion, tentative at first, developed into a grand rotation; like a ballerina pulling in her arms, the increasingly compact body developed a faster spin.

The heat produced by gravitational collapse built up at the center, while the nebula flattened like a ball of gelatin on a spinning turntable. The swirling disk was still as dark as a long-playing record, the lumps in its material slowly cooling to become the planets. One grand but unrecorded day—the real dawn of the solar system—the central ball's core became hot enough for nuclear fusion to begin, and this birth of the sun blew away the surrounding embryonic fluid like a fan clearing the neighborhood of smoke. The planetary lumps remained in the grasp of the sun's gravity, paying it unhurried obeisance with a promise of eternal revolution.

Thus our common origins transcend even the most exalted individual's family tree. Walking about the streets of our cities and living next door, other fragments of the same departed sun go on errands or change their children's diapers. Their every act is a confirmation that the unknown blue sun's death did not pass quite without consequence. We are made of stardust.

Now, looking toward Orion, our eyes, built of star fragments, witness the process afresh. The nebula's baby suns, too newly born for any of their planets to have had time to cool and become comfortable, are not the place to look for life—yet. This is still the early stage, the chrysalis of the unknown. If the mysterious unfolding of life is to occur, we must come back in another billion years for a further peek.

For the moment it is enough to see the lovely campfires scattered like an army encampment in the Orion cloud, and to revel in the ineffable feeling that comes from attending a birth. Peering at Orion, and having contemplated our own local genesis, we could even be inspired to delve into one of cosmology's major goals: the understanding of the creation of everything.

This ambitious leap, however, puts us in danger of awakening arguments long used by philosophy, religion, and metaphysics. For example, envisioning a "beginning" invites the question: What existed one minute earlier? The answer "Nothing" has never been satisfactory, for it's uncertain that there can even "be" nothing. (The verb *to be* flatly contradicts nothingness.) So, yes, semantics gets pulled into it, as well.

Each discipline will have its own answer, or method of evasion. Metaphysicians would argue that since eternity or timelessness is not graspable with the rational mind—nor is the concept of something existing "before the beginning"—the question of the universe's birth is insoluble with any dualistic mechanism, including rationality and science.

Maybe so. But our province, and one of the assignments of cosmology, is to come up with something at least provisionally acceptable, and we can accomplish this with a few simple clicks on the keyboard. We insert a tidy qualifier. We speak of the origin of the *present* universe.

"Present universe" simply means: *that which started with the Big Bang and is now expanding in all directions.* For if, as now seems certain, all clusters of galaxies are racing away from one another, it's a simple matter to trace their paths backward to find their location at each point in the past. Such extrapolation clearly shows that everything occupied the same spot at the same time, some 15 billion years ago.

Present knowledge offers only guesses as to why this unimaginably compact embryonic universe started expanding. But its first moment of existence as an ultradense egg can, in a real sense, be considered the nativity of the present universe. Perhaps it "got here" by collapsing from a previous existence. Maybe the universe breathes in and out every 80 to

100 billion years like the old Hindu legend of the "breaths of Brahma." Such an oscillating universe could allow new laws of physics each time, new adventures.

One thing is certain: We can find no trace of any previous universe because the Big Bang's unimaginable temperature and density pressure-cooked everything, and then some. Not one fragment of an atomic particle can have survived from the BeforeRealm to provide the faintest clue. The delete button was pushed, the machine reformatted, melted, vaporized, and reconstructed. No files are retrievable.

Some believe that the cosmos, dynamic as it is, lives in a sort of "steady state" of re-evolution that has no real birth or death. If so, the question of a beginning is meaningless. But many cosmologists disagree with this view, and their strongest argument invokes the concept of entropy.

According to most physicists, the universe is continually winding down from order toward chaos, from structure to smoothness, from energy flow toward quiet equilibrium. This widely accepted notion of entropy increase can be observed in everyday life.

Throw some ice cubes into the bathtub. When you return later, you'll see the result of entropy's cornerstone, the second law of thermodynamics. The cubes have melted; the water's reached a stable, unchanging temperature. You started with order, structure, and separateness. You ended up with smoothness, with no movement from one area to another, no further evolution. This point of inertness, in which energy transfer no longer occurs, can be considered a kind of death.

Is this the ultimate fate of the universe itself?

Physicists who subscribe to the standard interpretation of thermodynamics' second law—and nearly all of them do—believe that such a bleak view is realistic. They point out that order always tends to degenerate into disorder, like the drawer in which you keep your underwear. It can't be any other way in a closed system, one without outside influence. And isn't the entire universe, after all, a closed system?

This view supports the Big Bang and the existence of the passage of time. It says, in effect, that we've devolved from a past era when things were more highly ordered, to find ourselves at the present state of dissipation. This flow of events provides the only strong support for the presence of time; it suggests both a beginning and an end. The past was the

Big Bang. The present is the modern epoch of stars spewing heat into their frigid environs. The process ultimately arrives at a stagnant equilibrium: the future, the direction *away* from the Big Bang. It is not reversible.

It's hardly amusing—a future where the cosmos grows larger but less creative, like the federal government. Eventually a uniformity occurs where no energy flows. No new suns are born. Everything is unvaryingly smooth and cold. The only motion is the dark inertial tumbling of the frozen embers of dead galaxies, hurtling away from one another into ever more isolated emptiness.

This dark scenario is the most likely portrait of the future based on present knowledge. But scientific loopholes abound, driven perhaps in part by a philosophical aversion to thoughts of ultimate oblivion. Theoreticians are still human, and many share the view of optimists and romantics that the universe isn't so utterly pointless as to devolve into a dead and frozen emptiness. Indeed, one could make an argument based somewhere between science and philosophy that nature is demonstrably adept at creating new, surprising, even counterintuitive scenarios; we could be overlooking any of a thousand factors that alter our presumed destiny by 180 degrees.

We don't have to look far. Nobel Prize winner Ilya Prigogine pointed out over a decade ago that theoreticians may be applying thermodynamics' second law too broadly. It doesn't sufficiently consider such dynamic processes as electromagnetism and gravity, he said. It may work only in limited applications, such as our uninviting bathtub with the melting ice.

In fact, he believes, the universe is not decaying or winding down at all. Quite the contrary, it's evolving toward more structure and increased complexity! Certainly our planet's own experience is one of growing intricacy. A glance out your front door shows that the neighborhood is far more complex than it was in prehistoric times. Sophistication has increased, not diminished, to the point where plants and animals, mailboxes and automobiles have replaced the simple molten blob that was your driveway 4 billion years ago.

True, entropy applies only to closed systems without outside influence, unlike Earth which gets energy from the sun. Still, our situation may be analogous to the interactive way in which nature operates, re-

Is physical reality as straightforward as we assume?

minding us that, on the largest scale, the overall dynamics of the cosmos remain complex and poorly understood. It may be inappropriate to consider the entire universe as an isolated machine, and then expect it to behave as a steam engine does. We simply do not know enough to generalize so broadly.

This hopeful view represents a minority opinion among cosmologists today, many of whom would lay odds that a lonely kingdom of ice is in the cards. Moreover, in the inflationary model of the Big Bang theory, the curious balance between the observed expansion of the universe and the gravitational deceleration that's been slowing things down since day one is accounted for. If it's true, the universe will forever expand, but at an ever slower rate.

Still, most cosmologists would also concede that their knowledge, still in its adolescence, is expanding almost as rapidly as their subject, with inevitable surprises in store.

In any event, if the universe is really running down entropically toward disorder, it must have been *more* structured initially. Was it? And if so, how did it obtain such a high degree of order in the first place? Again, consideration of endings carries us to the beginning.

A fairly high initial order comes with the territory if the universe began as a "quantum fluctuation," a larger version of events believed commonplace in the subatomic realm. We could call it the "free lunch" hypothesis. It stems from the idea that particle-antiparticle pairs simply pop into existence all the time, only to vanish again spontaneously. The emergence of enough material to create a bonafide pair of universes would be extremely rare, but presumably the pre-universe universe had lots of time on its hands. Given enough of it, a massive collection of newborn thinglets could materialize: our universe! This even supplies us with *another* potential universe, the other member of our pair, an anti-universe born at the same time! It would presumably have opposite qualities to ours, such as tasty supermarket tomatoes and fathomable VCR controls.

"Quantum fluctuation" sounds impressive and is currently a popular explanation for the universe's genesis. Moreover, it accounts for the curious balance between the universe's expansion and the gravitational urge to collapse. But that notion of "something arising from nothing" remains physics-speak to many. After all, whether the universe did begin at a fixed point in time or whether it has always existed as a birthlessly infinite item—either way, its *ultimate* source remains a mystery fully as profound as when the first humans considered the question. When all the onion layers have been peeled away, science's careful fingers may find themselves grasping not only empty air, but a vacuum inhabiting the wrong dimension. Like someone trying to use a screwdriver's handle as a hammer, our present frustration may stem from employing the wrong tool for the job. The ultimate genesis of the universe may well demand a level of awareness entirely beyond mathematics, and perhaps even outside the aegis of linear thought.

This is not anti-intellectual. It is a realistic humility born of the improbability that our way of thinking can imbibe every morsel of universal reality. Still, we're not about to jump ship: It was science that brought us to this point in the first place. Our very knowledge of the night's nurseries and the celestial birth surrounding us in space was discovered through the ability to construct telescopes and then to comprehend what was seen. It has enabled us to raise the Orion and its myriad sister nebulas from their medieval designation of "cloud" to today's understanding that here, before our eyes, is nothing less than the birthplace of planets and suns!

As the millennium changes, we thus have a good knowledge of local birth, but have not yet been granted a reasonably secure awareness of the genesis of the cosmos. We grasp the parts while the whole still eludes us, and cannot say whether the regional process is mirrored on the largest scale.

And while such questions challenge our brightest minds, Orion crosses the icy night, a cosmic cradle bestowing the awesome breath of creation on the next generation of our galaxy's life.

Moonstruck

The moon influences mental health. The moon pulls on our bodily fluids. The moon is larger when it's setting. The moon governs menstrual cycles. The moon can affect the weather. . . .

If all the moon's alleged powers were real, we'd be a species of lunatics controlled from outer space. Fortunately, few such lunar beliefs are factual. (Of those cited, only the last is valid.) Yet the moon's real personality is amazing enough, and includes some surprising talents as powerful as the ancient myths they've displaced.

The moon's dominance of the night sky, and the fact that it's *the only nocturnal object most people can identify,* has always granted it an importance disproportionate to its modest size—one quarter of Earth's. Like a charismatic actor whose opinions gain more attention than they merit, our brilliant moon is endowed with a perpetually exalted position in poetry and folklore.

In reality, our moon would hardly rate a second glance almost anywhere else in the universe. Though a substantial satellite, it's just not a very impressive celestial body, with an uninteresting gray color, a long-dead geology, and a surface utterly without life or animation. When NASA planners contemplate manned missions, nobody talks about colonizing the moon.

This may seem surprising in light of Neil Armstrong's "one small step" utterance from the lunar surface, which reinforced expectations of repeated exploration. But the moon is so devoid of valuable assets, and so empty of the true gold sought by colonizers—water—that moon missions are essentially exercises in fantastically expensive water hauling. Had the lunar rocks not been found to be uniformly anhydrous (meaning bone-dry—you can't even squeeze water out of them chemically) and if humans didn't need prodigious (and, at eight pounds per gallon, weighty) quantities of it for survival, that first invasion in 1969 might truly have been one small step toward a permanent honeymoon there.

Incidentally, let's set the record straight about that historic little speech. The actual first words from the moon were *not* as they are now recorded. Books quote Armstrong as saying, "That's one small step for a man, one giant leap for mankind." Aside from the fact that "mankind" is now a bit too gender-specific for many people's tastes, it's a fine sentiment, a good speech.

However, he didn't say that.

Get hold of an actual tape from July 20, 1969, and you'll hear him say, "That's one small step for man, one giant leap for mankind." No "a" before "man." This one small omission by the man produced one giant leap of confusion for mankind—but not initially. As those first words were broad-

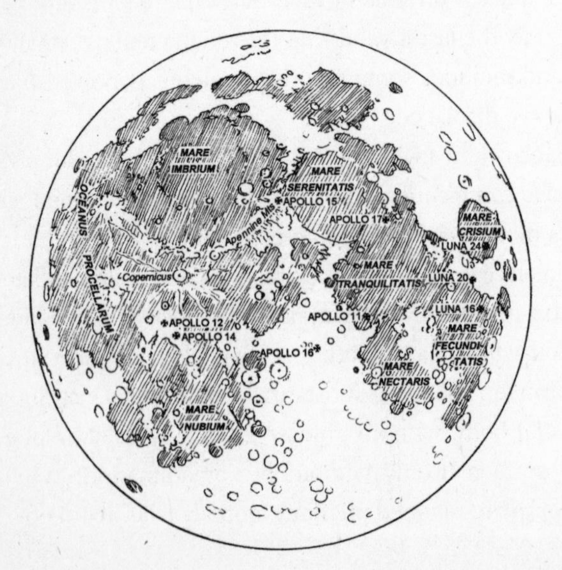

Human impact on the moon: The six Apollo *vehicles were manned; the Russian* Luna *landers were not.*

One small step for man . . . or was it a man?

cast from the moon, everyone cheered as indeed they would have had he instead sung "Moonlight Serenade," or ordered a pizza, or anything whatsoever. Few who applauded paused, like the boy inspecting the emperor's new outfit, to ask: What's that supposed to mean? "Man" in this context apparently meant "mankind," but then the two parts of the phrase contradict each other. It wasn't poetic license. It wasn't figurative speech. It was unintelligible!

No commentators, in the midst of that triumphant frenzy, questioned the emperor. After all, a mistake seemed inconceivable. Armstrong had had months to plan his immortal words. NASA claimed it had no hand in their choosing; the astronaut had been free to say what he wished. Perhaps he'd rehearsed it too often to himself, the way our overplanned speech to the boss or fiancée never quite works out. It occurred to no one that he'd come up with jabberwocky.

There were no official errata. Just simply, quietly, the first letter of the alphabet was added to the first words on the moon, and that's how history was made. Or unmade.

The second person's first step on the moon, taken by Buzz Aldrin twenty minutes after Armstrong, didn't go much better. Just as Aldrin placed his foot on the lunar surface, his urine bag, strapped inside that boot, broke. His first step—and every step afterward—could have been characterized as "one small squish for a man. . . ."

While such missteps were a natural part of our initial human encounters with the moon, the greatest leap of lunar knowledge occurred 359 years earlier, when the first crude telescope was pointed its way. Galileo's startling discoveries in January 1610 can be duplicated today by anyone with ordinary binoculars. In truth, the cheapest field glasses are far clearer than Galileo's early device, which produced blurry images unintentionally fringed with psychedelic color. Anyone who assumes that "mere binoculars" are inadequate for providing an immediate thrill (without resorting to the neighbor's window) has never observed the half-moon through them. A few days on either side of that phase, when sunlight optimally illuminates the lunar surface, any magnification yields dazzling results.

The cheapest department store telescope then supplies the owner with intricate lunar discoveries such as the ever-changing shadowing on terraced craters and mountain peaks. The moon is as easy and satisfying as it seems. Its features even maintain the same orientation toward us, thanks to its apparent lack of rotation.

"Apparent" because if the moon really didn't spin at all, it would present first one side and then the opposite hemisphere as it circled our planet in its monthly orbit. But it *does* rotate, in the same period as its revolution, so that, like a watchful lion tamer circling the cage but always facing the beast, the moon keeps the same face eternally earthward. This kind of synchronous rotation has no mystical significance; Earth's strong tidal pull has locked the moon's spin as if a cord were attached from our planet to its nearest surface. This phenomenon is common in the satellite systems of other planets, too.

It means the familiar markings on the moon do not change significantly, a curious fact thoroughly taken for granted. It would be as though, when watching a person approach, pass, and then walk away from us, we kept seeing only her front side. This constant viewing of the same hemisphere allows people to construct unchanging Rorschach shapes out of the reliable dark spots that are the lunar "seas" (areas of solidified lava). The major dark blotch on the right, for instance, is the Sea of Tranquillity,

Low-power view of the moon near first quarter

Medium-power view, same night, showing detail visible in any small telescope
PHOTOGRAPHS BY THE AUTHOR

The full moon through binoculars

It's easier to see a woman than a man in the moon.

where that first speech was delivered. The *seeming* lack of rotation means the rubbish left behind by the *Apollo* crew will sit forever in that same right-hand corner of the moon as you observe it.

The frozen catatonic grin of the moon's face makes for easy learning of the lunar topography. A single paragraph can summarize the most striking binocular features revealed when the moon is optimally illuminated—as a half-moon, or a bit fatter:

The mountain range near the middle, whose jagged peaks tower straight up at you like sharks' teeth, are the lunar Apennines, their inky shadows cascading onto the surrounding plain. The crater near the center with the spectacular terraced walls is Copernicus, best seen two or three days after the first quarter phase. Like bubbles in boiling oatmeal, a riot of endless craters dominate the southern region; they're just some of the thirty thousand craters visible from Earth, all caused by ancient meteor impacts. Dark, silky-smooth plains lie in vast blotches; these are the **maria,** the "seas."

The detail vanishes at full moon, when sunlight shines straight down to whitewash all shadows to oblivion. Then every crater and mountain seems to disappear. Only raw beginners imagine that a full moon offers a tempting telescope target. Astronomers greet its monthly sky-ruining arrival with the same enthusiasm the rest of us reserve for an IRS audit. During that phase, only the dark blotches (the maria) predominate. They're what form the traditional "man in the moon" seen with the naked eye—although a woman in profile is much more apparent (and these are modern times).

Be forewarned: You are about to be hexed. As the illustrations on the facing page confirm, an obvious woman (complete with an appropriately archaic hairdo) is impossible *not* to see once you've noticed her the first time. For the rest of your life, particularly if the moon is near full, and especially through binoculars, you cannot see the moon without perceiving this woman. She will haunt you forever. Inspect the illustrations at your own risk.

There is no risk, however, in gazing at the moon through binoculars or a telescope of any size, even when it's full. While larger telescopes can increase the moon's brightness several thousandfold (to the point where the observer sees a blotchy after-image upon turning away from the eyepiece), the moon's asphalt-dark surface, which reflects only 12 percent of the light striking it, can never pose a danger to the eye despite being, in essence, sunlight.

The moon provides more telescopic detail to the amateur observer than anything else in the universe. Indeed, the telescope's contribution to lunar knowledge was so hard to top that nothing surpassed it for three and a half centuries after Galileo. It finally happened in 1959 when the Russian spacecraft *Luna* 3 sent back the first pictures of the moon's far side.

One often hears of the moon's *dark* side, but there's no such thing, any more than Earth has a permanent dark side. All parts of the moon get two weeks of sunshine and then an equal duration of night, as sunrise creeps along its surface at under ten miles an hour. (A runner with enough stamina could keep nightfall at bay!) However, there *are* near and far sides to the moon so far as Earth is concerned.

Everyone assumed through the centuries that the hidden side of the moon would be similar to the hemisphere facing us. Of course, prior to 1959 nobody knew for sure. It was *possible* that the moon was actually a half-shell, a facade supported from behind by scaffolding made of two-by-fours. This, however, was deemed unlikely. (Impossible, actually, since a bit more than half the moon can be seen over time, in a wobble called libration that lets us peek around the edges a bit.) But nobody was prepared for the utterly distinct surface of the hidden hemisphere.

The first pictures were a shock. The two sides are so different, they're almost visions of separate worlds. The back side, undoubtedly because of scantier tidal stresses induced by Earth, has had less vulcanism in its distant past and consequently far fewer of the dark blotches that are so prominent on the side we see. Conversely, ancient craters buried by lava on the hemisphere facing us remain unscathed on the far side, which explains why that part of the moon is much more heavily cratered.

The Soviets wasted no time putting Russian labels on every conceivable feature there, in a wild orgy of name-dropping. To this day, that all-Russian hemisphere remains a quiet embarrassment in U.S. textbooks, mitigated only by the fact that nobody on Earth will ever see any of it.

The pace of new surprises started to quicken as landings by robots, and then astronauts, showed us that centuries of artists' drawings of its surface—based on logic and telescope observations—were wrong. The sharp pointy mountains we were sure dominated the moon's airless, waterless, erosionless landscape turned out to be as rounded as the hills of Scotland. Apparently tiny meteors, pounding the surface like a trillion hammers of Thor, had wrought their own form of erosion over the ages. They had also created the strange talcum-fine soil that clung to the astronauts and everything they brought with them. The depth of this soil had already been probed by robot landers, and determined to be safe for human trekking. Several inches deep, this baby powder crunching underfoot made strolls on the moon like hikes through the finest lakeside silt. Each

The far side: never seen from Earth

The near side: smoother, less cratered

Footprints on the moon will remain long after the human race is gone.

footstep left deep impressions that will be more enduring, on the moon's windless surface, than the greatest human monuments.

The Egyptian pyramids could turn to dust, be rebuilt, and turn to dust again ten thousand times over before these footprints vanish. The imprints of the dozen astronauts who walked the moon are steps into eternity beyond any made since life emerged from primordial ooze. They will survive the human race.

Equally enduring are the gold-plated plaques left on the lunar surface by the Apollo crews. At the insistence of the White House, the then-president's name and signature appear prominently on each. Should aliens somehow stumble upon our corner of the cosmos half a billion years hence, when *Homo sapiens* and all our traces are gone (or we have benignly evolved into something that would be unrecognizable to us today), the plaques and their inscriptions will still be legible. And if the extraterrestrials somehow decipher the recurring letters as a human name, they might wonder: What glorious achievement caused this single person to be chosen for immortalization beyond the species itself?

The name that will survive the human race? Richard M. Nixon.

If only one of the former president's famous quotes could have been included as well, such as "I am not a crook," it would add the final touch. Perhaps someday a return mission can remedy this.

(Other names, though not recurring, do appear on the plaques and emblems left on the surface: the names of people visiting the moon, and of the astronauts of both the United States and the USSR who died during the space program. The expression "we came in peace" appears prominently, an inscription that itself will likely be left in peace, undisturbed and unchanged, for at least one hundred thousand times longer than the interval that has now elapsed since the fall of the Roman Empire.)

Another surprise reached us through signals broadcast from seismometers left behind on the surface. These instruments were designed to detect moonquakes, but registered their largest readings when the unneeded ascent stage of the lunar module crashed to the moon after the astronauts had successfully returned to the command module that took them back to Earth. With each impact, the moon shook . . . and kept shaking. The moon rings like a bell!

Unlike quakes on Earth, which dampen out in a few minutes at most, the moon's solid core apparently allows vibrations to echo back and forth like a colossal Oriental gong. The moon "rang" for 1½ to 3 hours each time it was struck. If only space weren't airless, we might even have been able to *hear* the moon!

Among the biggest lunar misconceptions are the parables about its supposed affinity for water, an interesting myth considering its utter dryness. But the mistake is understandable since beachgoers see the tides responding to the moon's invisible fingers and reason, "If the moon can pull the mighty oceans, and if my body is mostly water, then why shouldn't the moon affect me?"

This faulty logic also applies to menstrual cycles. However, the moon's 29.5 day sequence of phases has never marched in step with the majority of women. Moreover, only one other mammal, the opossum, has a similar cycle. If the moon really exerts an influence in this regard, the dubious honor is withheld from the entire rest of the animal kingdom.

All such misconceptions arise because of the confusion between gravitational and tidal effects. Obviously, they're not the same. Our planet orbits the sun, not the moon, illustrating the fact that the sun has the stronger gravitational pull on us. Yet the ocean tides primarily follow the movements of the moon. How can one body have the greater gravitational influence on us, yet another produce a stronger tidal pull? In our educated society, not one person in a thousand could explain this, if their lives depended on the answer.

Yet the answer is easy. Gravity falls off very rapidly with increasing distance, so that the moon's pull on the side of Earth facing it is stronger than its pull on the other side, 8,000 miles farther away. This difference is not what produces the tidal effect; it *is* the tidal effect.

The sun's much greater distance means that the difference between its pull on the two sides of Earth is less extreme. That's the entire story.

The oceans, being interconnected and fluid, respond to the moon's differential pull with a daily rise and fall averaging three feet worldwide, with many local variations having to do with resonances, the shape of bays, and numerous other factors.

People, on the other hand, average only five to six feet in length, so the moon's pull on anyone's entire body is even and identical. You'd have to grow thousands of miles tall before tides would resonate within you. So far as the moon is concerned, your present stature makes you a sizeless point.

The pull on your head and feet being identical, your bodily fluids have no inclination to migrate from their customary positions, any more than tea in a cup tries to climb up the sides when the moon passes overhead.

Nor does the moon particularly care about liquids. The solid body of Earth responds to the tidal effect with a daily eight-inch deformity of its own. (Which, incidentally, is remarkably small. That our planet can hold its shape so perfectly despite the presence of a nearby body with one-quarter its diameter is one of the structural success stories of the solar system. We are very well constructed.)

Even the air is influenced tidally. The difference between the moon's pull on Earth's near and far sides sets up a recently discovered atmospheric tide that generates a slight correlation between cloudy weather and lunar phase.

Perhaps, then, we've come full circle. While studies looking at admissions to mental hospitals show no link to the moon's cycles, sunshine *is* linked to mood. Thus do we finally have a mechanism by which the moon can affect emotion, or at least one's disposition—through its influence on whether the day is gloomy or sunny! (The sunshine effect, however, is very weak and shows up only statistically.)

Over the years, nurses and physicians in my astronomy classes have consistently claimed a link between births and the full moon. Many declared they were sure, having observed it firsthand, that births increase

Centuries of telescopic observation led us to believe that lunar mountains are sharp . . .

. . . but when we went there, we learned otherwise.

during that lunar phase. This contemporary bit of lunar lore has been noted by others as well, and is fascinating.

Because—it isn't true. If it were, we would all have learned in Biology 101 how the human reproductive system marches to a lunar rhythm. Instead, records and insurance company data show that births are dispersed evenly throughout the lunar cycle. (There is, however, a link with *season*. In the United States and a few other countries, births increase in September, perhaps signifying somewhat of a human fertility cycle—or possibly suggesting a bit more snuggling during the long, cold December nights. But there is no link to the moon.)

Moreover, if births increased at full moon, they'd then decrease at other phases, and the human population would have birthdays showing a reliable up-and-down sine wave figure each month, a pattern that does not exist. All this only makes the problem more interesting, for the real question then becomes: Why do so many *perceive* a birth increase if it isn't so? A persistent myth is, after all, fascinating in its own right.

Here there are no scientific data, but I'm convinced the answer lies in this scenario:

En route to work, the medical professional notices that the moon is full. (Most people will call the moon "full" even when it is out-of-round, so that a full moon can occur on any of several days each month.) Now if, by chance, there does happen to be an unusually large number of births that night with an attendant hectic work pace, the medical staff will be abuzz with "Aha, I knew it, it's a full moon!" and the link becomes strengthened. If, however, the night produces a normal or subnormal number of births, staff members do not think about the matter at all, for there is nothing exceptional to attract attention to the issue. Similarly, if a high-birth night occurs when a full moon has not been noticed, again no connection is made or noted. The preferential reinforcement is always toward a full-moon/high-birth-rate correlation, without the individual's awareness of the process.

I'm convinced this accounts not only for the spurious lunar/birth link among medical personnel, but many other misconceptions about the moon that have persisted through the centuries.

This does not mean that the moon is powerless to affect our lives. Its influence on marine life is undisputed, and there may be many other effects yet undiscovered. For instance, one recent study found that earthquakes have a small but distinct correlation to lunar phase.

This makes sense. When the moon ventures near Earth in its elliptical orbit, tidal stresses increase. Stresses also rise during the new and full phases when solar and lunar effects work together. When both these influences coincide, as happens once or twice each year, our planet is subjected to unusually strong (so-called **proxigean**) tidal forces that can wreak erosional havoc on beaches and penetrate Earth's crust. Such torques are relatively small (less than 0.1 percent) compared to the stresses normally found in crustal rock. But if an earthquake is on the edge of occurring, they can supply the last straw. We'd thus expect most earthquakes to commence independently of the moon's position, but we'd also expect their overall onset to deviate just a little from absolute statistical randomness. In short, we'd envision a small correlation between earthquakes and the moon—and that's just what we find.

Lunar discoveries continue, even for the casual skygazer. The naked-eye moon watcher can observe the different sizes of the moon as it wanders nearer and farther from us in its elliptical orbit. Its varying brightnesses are also obvious (it's more brilliant when it's higher, when it's nearer, and in winter when sunlight striking it is 7 percent stronger—since the Earth-moon system is several million miles closer to the sun at that time of year).

Then there are the strange effects. One is the famous **moon illusion,** which makes a rising or setting moon appear enormous (see page 270). Another is the way the full moon looks flat, like a dish, rather than three-dimensional. When any sphere is illuminated straight on, its center is normally brighter while the encircling edge (limb) is darker (see color plate 3). This is caused by the light's low angle at the edges, and gives a definite sense of solidity or dimensionality to the object. But the moon looks weirdly flat when it's full, as if it were a disk *painted* onto the sky.

Noticed since the Middle Ages, this curious effect is now known to be caused by the moon's rough, powdery surface, which bounces incoming sunlight in all directions and prevents it from reflecting light the way a smooth or gaseous body would.

Such oddities, observed since before Galileo's day, are as revelatory to each new generation of sky watchers as they were in bygone centuries. Until the moon's expected breakup into a stunning Saturn-like ring several billion years from now, we will never lose our fascination with the dark gray sphere that silently accompanies us in our odyssey through space.

Fading to Black

We who live in the Northern Hemisphere face our greatest darkness from November to January. In most of Europe, Canada, and the United States, the sun shines less than seven hours a day, producing a murkiness in those months that, to no one's surprise, is linked with depression. But such long nights are a perfect time to explore the shadowy topic of blackness.

You'd think a subject as basic as darkness would be as simple as black and white. In reality, true blackness scarcely exists in our neck of the cosmic woods, and about the only way to experience it is to lock yourself in a light-tight closet. As we'll see, a search for darkness and its shady phenomena will take us to some of the strangest places in the universe.

Certainly, night itself is not black.

The annual 10 percent increase in artificial lighting that occurred during much of the past three decades—the result of added streetlights, billboards, malls, parking lots, and yard lights—has given modern skies a permanent artificial glow. A peek through a five-dollar spectroscope (an instrument every science-oriented person would enjoy) dramatically reveals this light's composition as a pattern of specific colors emitted by streetlights' glowing gases. But even the unaided eye sees plainly enough that the urban night sky radiates an eerie presence whose strange tint exists nowhere else in the known universe.

A quarter century ago this synthetic skyglow was a steely gray-blue, a result of the widespread use of mercury-vapor streetlights. Those intense cold-hued lamps shine by a more efficient process than the simple incandescent bulbs they replaced. Instead of using a filament heated white-hot by an electric current, a process unchanged since the nineteenth century, the newer streetlamps force a much higher voltage through gaseous mercury, exciting its atoms to a higher energy level. As they spontaneously return to their preferred, lower state, they emit photons (packets of light). By the 1960's, city streets throughout the United States were illuminated by mercury lamps, and their ghoulish blue-white presence became the nocturnal norm.

When sodium rather than mercury vapor is used, less electricity is needed and a more pleasant color is produced. It was thus only a matter of time before the tint of the night would alter; the energy-consciousness of the 1970's helped accelerate this switch to the pinkish-yellow sodium-vapor lamps that are now the standard.

Streetlights are generally replaced only as they burn out, so mercury sanctuaries will continue to be common for years to come, providing bluish oases for our inspection and comparison. But the sodium era is firmly established, and explains the amber umbrella that now floats over most cities. It's almost as if humans changed the color of the night to celebrate the dawn of the new millennium.

Major metropolitan centers announce their presence from as far as 60 miles away with this saffron reflection from overhead clouds and atmospheric dust. The brightness of this luminous sodium-induced blanket is impres-

The icy blue light of mercury-vapor lamps is still common, despite their energy inefficiency.

sive, although it would be fully appreciated only if the causative ground lights could somehow be masked. A power blackout would do the job nicely, except then the glow hovering above the city would itself vanish in a few microseconds (millionths of a second), the time lag needed for the light to cease reaching the clouds and bouncing back down to the observer's eyes.

It's fascinating to imagine a city's sudden electrical failure, followed by the sky's milky glow disappearing an eyeblink later. Blackouts of other celestial lights would produce even more dramatic delays. If the sun were to blow up, say, at noon, we wouldn't experience this unfortunate turn of events until 12:08, a reprieve granted by light's travel time to Earth. The full moon would vanish about three seconds later, since light passing us as the sun failed would still continue onward, then reflect back to our eyes from the lunar surface.

Farther out, Jupiter and Saturn would fade from our sight about two hours after the sun turned off, the outer planets Uranus and Neptune many hours later. How disconcerting to watch these lights of the night turning off one by one as the final photons of sunlight sped off into the distance like a receding train!

Of course, the past century's nocturnal transformation is less obvious in smaller cities requiring fewer lights than the major megalopolises. Still, population centers of more than one hundred thousand now sport skies that will fog (overexpose or saturate) a fast film in less than a minute and are often bright enough to read by!

Rural regions (particularly in the low-density areas of the western United States) do enjoy unpolluted skies, little changed from a century ago. These skies are natural—but still aren't black. A moonless night in a pristine region exhibits a firmament that is dark blue-gray, not black at all. The source of this illumination is neither the stars nor human interference. It's the sky itself!

Airglow is the name for the sky's natural fluorescence. It's caused by the sun's invisible but powerful ultraviolet radiation, which excites our atmosphere's gases to glow like the hand of a luminous watch. This background glow varies, but is usually as bright as the combined light of all the stars. That's why you can still manage to see well enough to walk safely—once your eyes are fully dark-adapted—even on the darkest country road,

A black cat illuminated by bright sunlight in light and dark environments

except where overhanging trees obstruct the sky. "Black as night" is not a particularly apt expression.

So if night is not truly dark, is asphalt? Onyx?

Neither! They only *seem* black relative to things that are more reflective. What appears black or white is merely the eye's subjective experience as it compares objects viewed under common lighting conditions. In bright light our eyes adjust to the intense surroundings by perceiving the most brilliant portions of the scene as "white," while assigning everything below a certain level of illumination to "black." However, a tremendous amount of light still comes at us from the "black" objects.

In a dimmer setting, our pupils as well as photochemical retinal changes reset the black and white levels. A less reflective object will appear consistently darker in every scene, but its perceived blackness remains subjective. Put it this way: A black object seen in a brightly lit environment would look dazzlingly white if it could be transported as is to murky surroundings. For example:

A black cat in sunlight is actually some two thousand times whiter and brighter than snow illuminated by the full moon.

If this sunlit "black" feline could prowl in a nocturnal setting, it would be blindingly, supernaturally white, like the cat that kept the mice out of the nuclear power plant. Its brilliance would cause surrounding objects to cast shadows.

Determining which is brighter—snow in moonlight versus asphalt in sunlight, or coal in sunlight versus an egg in candlelight—is pretty simple with a photometer, another device that belongs on the naturalist's wish list. But a camera's light meter will do fine. Either can demonstrate that nothing we encounter in an open setting is fully dark. It's obvious, then, that any serious exploration of blackness must carry us beyond Earth itself.

Here and there, the cosmos contains extremely shadowy objects that are popularly considered black. Well-known examples are the black dwarf stars. Far from rare curiosities, black dwarfs are the final stage of life for most stars, including our own sun some 10 billion years in the future. Merely the size of Earth, these collapsed stellar relics, so dense that each cupful outweighs a fully loaded bus, are probably strewn throughout the universe. Nobody knows for sure because they're invisible to our instruments. But while textbooks refer to them as "black," they certainly wouldn't appear absolutely dark in person. Their nuclear fires long dead and emitting no light, they nonetheless have no trouble reflecting some of the starlight striking them. Visit a black dwarf and you could grope along its surface just as we can walk down that dark country road—by light from the sky. Cold and murky, yes—but not black. Bring along firewood and you could camp out on a black dwarf.

(Think of it: sitting around a campfire on a star. An appealing image, except that the gravity would be so strong you'd never be able to lift a marshmallow to the fire. For that matter you also couldn't stand, sit, or even breathe. Not that breathing would be easy in any event, since there'd be no atmosphere. And without air, how much heat could that campfire give off? The whole idea is losing its charm. Cancel the trip.)

We have now entered a limitless celestial domain—the realm of objects so faint that they lie below the threshold of vision. We first ventured into this dusky kingdom late in 1609, when the first telescope was aimed skyward. To his surprise, Galileo saw a flood of stars that had been forever hiding from human eyes. Inadvertently, that baptismal moment of celestial exploration launched a perpetual myth. For most people believe now, as they did then, that a telescope's function is to make things bigger. Not so; its most important job is to make objects brighter.

Many celestial targets are already large enough for our inspection and need little or no magnification at all. The problem is simply that they're invisibly faint. A good example is the nearest major galaxy, Andromeda, which takes up as much space in our sky as a row of eight full moons. It requires very little magnification to be studied. Increased brightness is what's really needed. Even when we do want enlargement, the light must first be intensified to compensate for the fact that a magnified image is spread out and dimmed.

The most common misconception surrounding telescopes is the idea that magnifying power defines quality or worth. Highpoweritis is the disease of the neophyte. The fact is, astronomers using even the world's largest telescopes rarely employ a magnification greater than 300 times or so. More common is 50 times to 250 times, even among amateurs with very expensive equipment. By contrast, these same instruments amplify the brightness of the target at least two thousandfold. Telescopes are rheostats, light intensifiers. This is the primary quality that allows stars and galaxies to spring into view.

If you own binoculars, you can treat yourself to a quick and dramatic demonstration of the potency of *light-gathering power.* The best arena for this display is the night sky itself.

To the naked eye, fewer than three thousand stars can be seen on the clearest night—and that includes all the innumerable faint stars that crowd the rural sky. This always surprises people, since it's com-

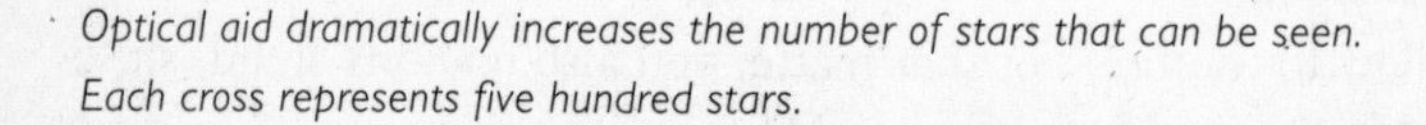

Optical aid dramatically increases the number of stars that can be seen. Each cross represents five hundred stars.

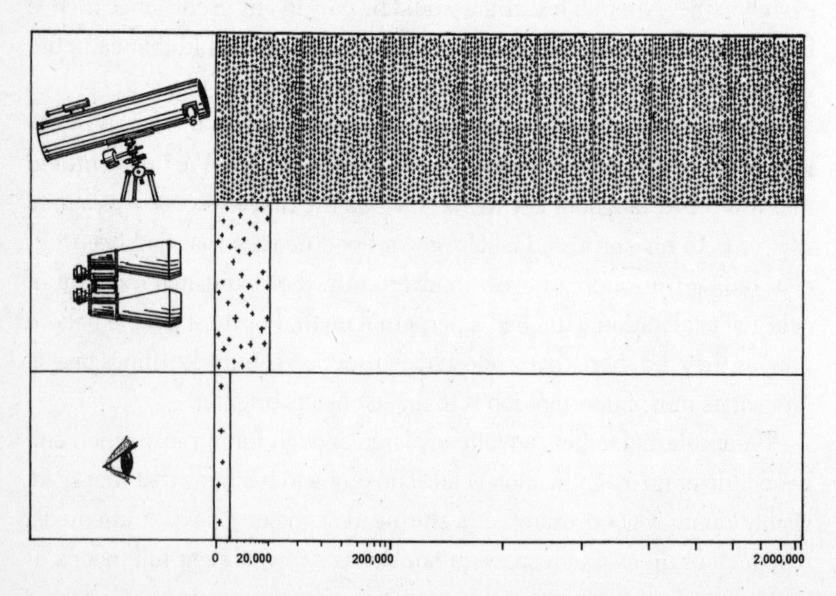

monly imagined that "millions" of stars flood the nightly heavens. *In reality, you could count every naked-eye star in under twenty minutes!* Now point binoculars skyward. For every star you could see before, at least eight new ones pop into view. The instrument's light-gathering ability (not its magnifying power) has magically materialized twenty-five thousand new stars.

The binoculars have also just disclosed a basic fact about the heavens, noticed since antiquity: There are few brilliant stars, many more medium ones, and a flood of faint stars. This hierarchy continues with a vengeance below the threshold of human vision. As telescopes get larger, the number of stars continues to increase geometrically. (But even the largest instruments detect less than a tenth of 1 percent of our galaxy's trillion suns.)

Still, today's marriage of ever larger telescopes with electronic amplifiers such as CCD's (charge-coupled devices) does an impressive job of plumbing and quantifying the murky depths below the limit of human vision. Recent advances have allowed telescopes to detect stars of magnitude 29—more than a billion times fainter than anything the unaided eye can perceive! This is very faint indeed: The light from such a star equals the glow of a single cigarette seen from 125,000 miles away, a dozen times farther than Washington, D.C., is from Tokyo.

Detecting objects of such breathtaking faintness allows us to peer 10 billion light-years into space, and brings up an interesting point: *To see these objects at all, they must be brighter than their background.* Therefore, the cosmic backdrop is darker still. Does this mean that space itself is absolutely black?

We usually picture deep space as the very paradigm of blackness, forgetting that, being a vacuum, space must actually be colorless as water. But were we to teleport ourselves to some distant realm far from our sun, enough light would still arrive from the rest of the universe to disqualify it from perfect inkiness. It is, to be sure, quite dark: The average brightness of the sky as seen above Earth's atmosphere is about a million times dimmer than this page appears if read by the light of a 100-watt bulb. It's the equivalent of the glow of a tenth-magnitude star (a star three times too faint for binoculars to detect) spread out over each moon-sized piece of sky.

But we don't need to venture all the way to the stars. The nearest and best-known "dark" object is the planet Pluto, whose very name comes from the *god of the underworld.*

Except, it isn't a planet. Sure, most people will keep calling it one, out of habit. But since the discovery in the 1980's that it's only half the size of the moon, we cannot really put it in the class of the other planets. It's more like an asteroid, or *minor planet,* and it even travels in an odd, eccentric, non-planetary-type orbit.

Nor is it an "it"—Pluto is a "they"! There are two of them! Pluto is a double object, two icy chunks orbiting each other every six days, the larger piece barely twice the size of the smaller. Perhaps Pluto represents the largest member of a new class of small asteroid-like objects that inhabit the outskirts of the solar system.

Misinformation about Pluto abounds. An oft-repeated error is found in such books as David Louis's *2,201 Fascinating Facts* (Crown): "To an observer standing on Pluto, the sun would appear no brighter than Venus appears in our evening sky."

This sort of thing has given Pluto an undeservedly dark reputation. But it's easy to calculate how much sunlight strikes its surface by using the **inverse-square law,** which keeps popping up in astronomy. (It applies to gravity as well as the intensity of light, and is simpler than it sounds: If any object—say, the sun—were to move three times closer to us than it is now, then both its brightness and its gravitational pull would become three squared, or nine times stronger.)

Pluto, on average, is 40 times farther from the sun than we are, so sunlight there is 40 times 40, or 1,600 times less bright than ours. This means the sun would appear to be magnitude –20, or over 1,000 times brighter than our full moon! Or, if you like, exactly a million times brighter than Venus.

Picture it: the brightness of a thousand moons radiating not from a disk—for the sun would appear too small from Pluto for the eye to resolve—but from a *point* of light. It would be blinding, dazzling. Objects would cast sharp shadows on Pluto's surface. Hardly the place to find blackness.

This Dracularian obsession with darkness is now driving us to scour the entire cosmos. Perhaps a black hole will reward us with the pure darkness we seek. Finally we've arrived at an object that neither emits nor reflects any light. None. Shine a searchlight at a black hole and the beam reveals nothing at all! Perfect blackness . . .

. . . maybe. Because there's still a qualification: Black holes appear black only when seen from the outside. Inside a black hole there may well be plenty of light, since a star collapsing sufficiently to achieve black-hole density can continue to emit energy, even though that light remains trapped within. Black holes have their own chapter, so for now let's just mention that the entire universe could be a colossal black hole; if so, there's obviously plenty of light all around us here on the inside.

Back on Earth, one striking indicator of the scarcity of blackness is the persistence of color vision. At night our eyes automatically shift from the color-sensitive cone-shaped cells to the retina's low-light machinery: rod cells whose only defect is color blindness. For you to perceive color, the ambient light must be at least 0.03 lamberts (a measure of illumination)—not bright, but not really dark, either. This is why artists portray nocturnal scenes absent of any bright color, and why groping around a dark attic or basement is an adventure in black and white. But a quick glance reveals that urban or suburban nighttime settings bring no cessation of color.

The right to a naturally dark nighttime is not one of the better-known environmental issues. Astronomy buffs, who have always held that "black is beautiful," howl in vain against upward-pointing billboard lights and other sky-brightening manifestations of "waste lighting." They point out, for example, that motion-detecting floodlights are more effective at deterring crime than steady glaring yard lights, do not pollute the sky because they're aimed downward, and shut themselves off automatically instead of remaining lit throughout the night. Municipalities and developers have similar choices. Energy and the night sky increasingly are being saved with the installation of recessed fixtures (instead of unshielded lamps), which aim all the light where it's needed: on the ground.

Most city dwellers are either oblivious or resigned to their loss of such dark-sky phenomena as the northern lights and Milky Way, reduced to mere museum curiosities synthetically displayed in planetariums. But it would be a needless misfortune for them to vanish from our rural skies as well.

The issue has never been a ghoulish aversion to light, but rather an appreciation of the natural phenomena that delicately materialize—like the magic of a Rembrandt—from the darkness.

Meanwhile, if you really want to see black . . . clear out some of that junk, and step into the closet.

Two-Dog Night

When early nightfall cloaks the dinner hour and barren trees bend in the chilly autumn wind, then begins the wintry reign of Sirius, the brightest star of the heavens.

The best known of all stars through the ages, it deserves its fame. While giant Canopus offers halfhearted competition in the tropics, there is no contest in the skies above Europe, Canada, or the United States: Sirius is four times brighter than its nearest rival.

No surprise, then, that many civilizations throughout history worshiped Sirius as a god, and acclaimed its rising and setting with monuments, or aligned pyramids or passageways to mark its highest nightly position. Known since antiquity as the Dog Star because of its prominence in Canis Major, the "Big Dog," it delivers enough lore and legend for twenty normal suns.

Today we explore Sirius with renewed awe because of the knowledge that it marches through space with a bizarre companion—the nearest example of a fantastically tiny ball of crushed nuclear fire.

Finding Sirius is never a problem, even if you cannot trace its constellation's pattern. Apparently the ancients had the same disability, for the people who drew the old star maps variously depicted the dazzling blue star as the dog's heart, paw, chest, or nose. Even today nobody can say with certainty which part of the canine anatomy Sirius represents.

It doesn't matter. Dominating the southern sky at midnight, the Dog Star can be outshone only by the planets Mars and Jupiter. And Mars appears so orange and Jupiter so creamy white that neither can be mistaken for Sirius' blue-white diamond dazzle. Moreover, Mars will surpass Sirius for only a few months in 2001, 2003, and 2005, when our worlds briefly brush past each other in the corridors of the night.

In short, we're always safe in identifying the brightest, bluest wintertime star as Sirius. If ironclad confirmation is needed, the three-stars-in-a-row of Orion's belt point eternally leftward and downward toward the Dog Star.

You are unlikely, upon spotting Sirius, to sacrifice a dog to mark the occasion, yet that is exactly how the ancient Romans reacted. Back then the

The stars of Orion are in the upper right corner, drawn so that their sizes show their relative brightness. The belt of Orion points roughly toward Sirius at lower left. Procyon, another bright star nearby, is at upper left.

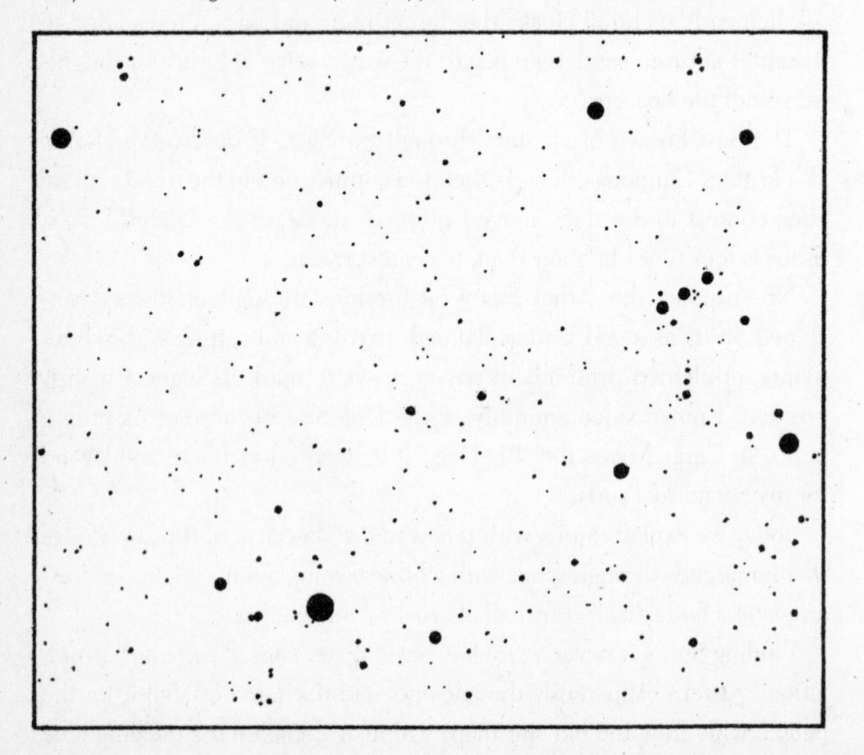

Egyptians welcomed the Dog Star's rising just before the Nile floods.

"influence" of Sirius was regarded as unlucky, even unhealthful, although this of course doesn't justify their hiring of canine hit men. Its lowly status was exemplified by a line from Virgil, who wrote that the "Dog Star saddens the sky with inauspicious light"—a view that persisted for millennia. Dante, obviously no patron of the Kennel Club, wrote of "the scourge of days canicular."

But in other cultures Sirius incurred quite a different reputation. In ancient Egypt, for example, Sirius was revered as a manifestation of Isis. Back in the second millennium B.C., when the constellations were oriented differently from today because of the slow wobble of Earth's axis, the Dog Star first appeared each year in late June, during the hot weather that preceded the Nile's flooding. Such an emergence was regarded as auspicious by the Egyptians, as it occurred just before the rains upon which their lives depended. Its godly position was further reinforced by the popular belief both there and in ancient Greece that the scorching weather was actually *caused* by Sirius, the result of an alliance between the Dog Star's dazzling light and the sun's rays. Even today we still use the expression *dog days* to mean sultry weather—though few realize that its origins echo distantly from those vanished centuries.

Sirius' ancient connection to heat may have produced the puzzling and persistent allusion, by people as diverse as Cicero, Horace, and Ptolemy, to its being a "red" star. That description has generated centuries of controversy, pitting scientists who maintain that stars cannot mutate in such

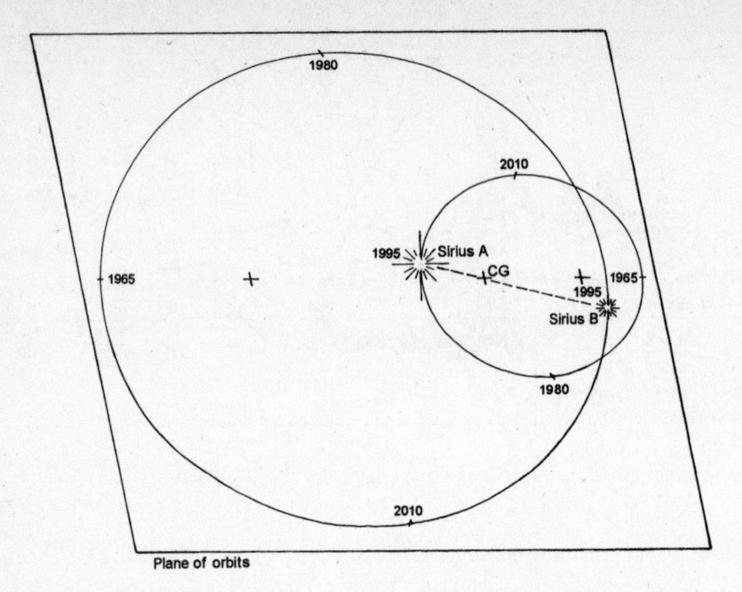

Sirius A and B revolve around a common center of gravity (CG). A is 2.5 times heavier than B, giving it a smaller orbit that's always more than twice as close to the CG as its white dwarf companion. (The other crosses show the remaining foci of the two stars' orbits.) Note the closeness of the pair in 1995 and the wide separation fifteen years later.

a brief cosmological instant against those who argue that so many early writers couldn't all be wrong. What else, they say, could Homer have meant when he compared Achilles' copper shield to the color of Sirius? Others, however, point out that Sirius often looks red when rising or setting, just as the moon and sun do, and it was at those times that rituals involving the Dog Star were performed.

Some have speculated that Sirius' companion star may have rapidly collapsed from a previous red-giant stage, where it threw ruddy brilliance into their combined dazzle. But astronomers generally agree that such a transformation requires far more time than the few millennia since the Roman Empire. And the ancient Greeks called other stars red that certainly were not warm-hued then or now.

In actuality, the rising of Sirius is often a psychedelic light show. When that dazzling starpoint shines through thick horizon air, it frequently appears as a twinkling kaleidoscope of rapidly changing color. Our observatory has received perhaps a half-dozen calls over the years from people

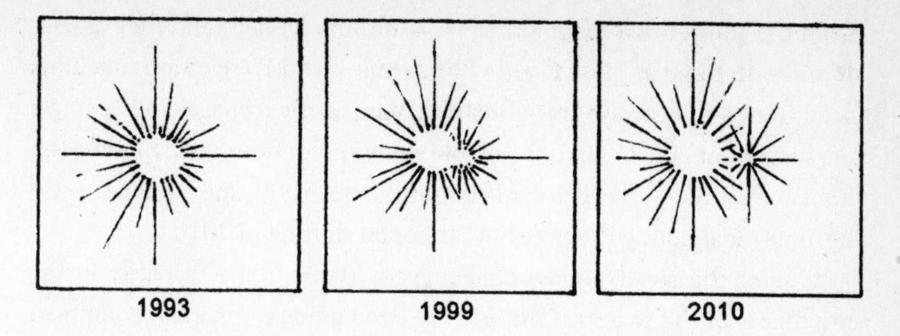

Separation of Sirius A (left) *and Sirius B*

reporting a "UFO" that proved upon investigation to be the madly mutating Dog Star. This common phenomenon occurs when the star's light is refracted by prismlike layers of air to transmit various parts of its multicolored spectrum. Even a brilliant green—a color never seen in stars—can briefly dominate, making the rising or setting of the Dog Star a memorable discolike event. As Sirius gains elevation it usually steadies to allow all its rays to reach our eyes simultaneously, producing its authentic blue-whiteness, a tint that signifies fierce temperatures twice those of our own sun.

When viewing Sirius we are actually seeing the combined light of two separate stars. Like the individual dots that make up a television image, these lie too closely together for the human eye to separate; the task is not easy even through a large telescope. The duality was first suspected in the 1830's, when careful measurements showed that Sirius wobbled drunkenly through space as if stumbling in the gravitational embrace of a fellow traveler. Calculations correctly determined the mass, distance, and fifty-year orbital period of the presumed companion. The renowned telescope maker Alvan Clark suddenly spotted the cryptic star in 1862 during the routine testing of a new instrument, then the largest in the world.

The difficulty is that the companion, Sirius B, is often lost in the dazzling glare of the intensely luminous primary, Sirius A. Steady atmospheric conditions, when stars aren't twinkling at the observer's site, are critical to their resolution.

The two are presently (in the middle 1990's) swinging through the part of their lopsided orbit where they're closest together. Although they are generally some 2 billion miles apart—equaling the distance of Uranus from the sun—that separation has now shrunk by half, and they appear from Earth to lie as close together as pennies whose edges are touching, seen from a mile away. This effectively merges their images into a single dazzling point of light, an unfavorable situation for observers. Finding the faint companion will get easier beginning about 1999, and conditions will become ideal during the twenty-year period starting in 2010.

One of the observational challenges is the extreme disparity in the brightness of the two stars. At first the companion's remarkable dimness caused bewilderment; with a high surface temperature it should radiate plenty of light. The explanation for the paradox is stunning: Sirius B is a tiny sphere—a star only about the size of Earth, yet with a mass some 350,000 times greater. It is dim because it has 7,000 times less surface than the sun.

The combination of toy size and large mass means that its material is packed to a density that challenges comprehension. A pailful of Sirius B weighs more than 6 million pounds, as much as the thirty-six-story Saturn 5 rocket that sent the astronauts to the moon. A cupful equals the weight of two cement trucks. Such extreme compression is hard to grasp; a lollipop constructed of Sirius B material would outweigh a car.

It is thus the great size difference between Sirius A and its diminutive sidekick that produces their remarkable dissimilarity in brightness. A emits the same light as twenty-three of our suns, but B feebly radiates only one five-hundredth the light of a single sun. Put another way, Sirius A is eleven thousand times brighter than its companion: a floodlight versus a candle.

Comparison of light from Sirius A and Sirius B. The cube at left has eleven thousand times the volume of the cube at right.

Gradually it dawned upon nineteenth-century astronomers that Sirius B, fantastic as it is, was no odd duck. As a white dwarf, it's a member of a surprisingly common fraternity of stars. Their dimness makes all but the very nearest hopelessly invisible, even through today's largest instruments. Nonetheless, over a thousand such stars have now been cataloged.

They are strange in every way. Their usual composition is pure hydrogen: no trace of any other element. With a gravity 150,000 times greater than the pull one experiences on Earth's surface, if you weigh 150 pounds here, you'd tip the scales at 22 million on Sirius B. Getting out of bed could be quite a task when you weigh more than a freight train. Yet moving around presents no problem to the star itself. A typical white dwarf rotates in just an hour (the sun, by comparison, spins in twenty-five days).

Eventually such balls of condensed fire cool enough to assume a solid structure. Indeed, they essentially become single crystals! Since the process takes about 10 billion years, these impenetrable floating jewels will be a common feature only in the old age of the Galaxy.

Of the fifty-eight stars that have now been identified within 15 light-years of Earth, five are white dwarfs. Such frequency is informative. Their abundance in our neighborhood implies they're common components of the Galaxy's population. The modern conclusion is that these crushed, ultradense stars represent a normal elderly stage of life for all stars that weigh about the same as the sun or less. It's now thought that 97 percent of all stars will eventually collapse into white dwarfs. When we look toward Sirius, we're seeing our own sun's future as a tiny ball of compressed fire.

Sirius B, like other white dwarfs, will survive virtually unchanged for billions of years, its slightest additional collapse translated instantly into new heat and light. But over time it will cool, becoming progressively redder and dimmer until a temperature is reached where one could theoretically land on its surface. Unfortunately, that would qualify as a serious lapse of judgment, since the white dwarf's original high density and overpowering gravitational field would remain. The ship and crew would succumb instantly to this viselike pull. Against such an awesome force one's motion would freeze like a frame in a jammed projector. Not a single breath could be taken, nor could one lift a finger to warn others. And a beacon pointed upward would have its beam reddened as its light's waves, struggling against the tug of gravity, stretched out and changed color.

Dogan mythology. How did an African tribe learn about Sirius B?

Such a bizarre **gravitational redshift**—predicted by Einstein in 1916—was confirmed by changes in the spectrum of white dwarf stars such as Sirius B. Simply put, the light loses energy in its upward fight against the star's gravity. While light's constant speed is preserved (the light cannot be slowed by the struggle) its waves are robbed of energy and lengthened, and thus appear redder. A few theoreticians have even tried, thus far unsuccessfully, to explain part or all of the nearly universal redshift of galaxies as such a gravitational effect instead of the customary interpretation, that the phenomenon results from the expansion of the universe.

Sirius B again took center stage in the late 1970's in, of all places, a book and series of popular articles dealing with the religious practices of the Dogan tribe of western Africa. That tribe, like cultures throughout history, worshiped Sirius, but with this astounding difference: Its rituals revealed not only knowledge of the existence of its white dwarf companion, but of Sirius B's orbital period as well!

Since that star cannot possibly be detected without the benefit of technology far beyond the tribe's capabilities, the often-repeated conclusion was that the tribe must have been visited and enlightened by extraterrestrials from the Sirius system.

"You can't be Sirius!" was the reaction from the astronomical community, which offered alternative explanations that attracted far less publicity than the original headlines about aliens in Africa. Skeptics suggested, for example, that the tribespeople had been visited decades earlier than thought, perhaps by missionaries who, upon learning of their interest in the brightest star, supplied them with information that eventually became part of their lore.

This minor debate exemplified the kind of improbable "controversy" found all too often in sensationalist journals. Few who embrace such wild speculation would likely employ such illogic in their own lives. If, for example, the family car fails to start in the morning, most of us would consider an empty gas tank or weak battery before entertaining the possibility of damage from an overnight meteor. But many seem unable to apply the same principles of probability to peculiar stories on the printed page. In this case, one could also argue the unlikelihood that celestial invaders would ignore the entire world in favor of a single tribe, before vanishing back into the night.

But if Sirians did pay us a call, they wouldn't have far to travel. Sirius is the closest star visible to the naked eye from most of the developed world. (From the tropics and Southern Hemisphere, Alpha Centauri can be seen; it lies exactly half the distance between us and Sirius.) At 8½ light-years, the Dog Star's sapphire light requires scarcely more than a two-term presidency to arrive through the vacuum of space. Its brilliance, then, has a dual origin. It really is a luminous sun, and it's the very closest that our eyes can behold.

If our galaxy were the size of North America, Sirius would sit just 500 yards away. The resplendent Dog and its amazing Pup are next-door neighbors and, barring a nearby supernova, which could temporarily usurp its dominance, will remain the night's most doggedly brilliant star for as long as human eyes gaze skyward.

Puzzles of the Polestar

No star is more famous, or easier to find—or the source of as much confusion—as Polaris, the North Star.

Lead a group of laypeople out under the night sky and show them Polaris, and the reaction is predictable. Their line: "But I thought the North Star was the brightest star in the sky!"

The North Star *is* amazing; it defies logic and probability. It begs to be understood and appreciated. But its brightness is not its calling card.

Of the six thousand stars visible to the naked eye, Polaris ranks a respectable—but no cigar—fiftieth in brightness. It's bright enough to appear even in polluted urban skies, but not brilliant enough to catch one's eye. It's second magnitude—a medium star, no great luminary.

No, what makes it notable is that it's the only thing in the sky that, from our earthbound perspective, doesn't move. Glance upward, and then return to the sky an hour later. Everything's shifted. Moon, stars, planets—all have migrated westward. The relentless turning of our planet has made western objects sink lower or even vanish below the horizon, while the eastern sky, that eternal theater of newness, now offers stars that weren't there before. In the south the celestial scenery has jumped rightward by three arm's-length fists every two hours.

But the North Star stays put. If, from your home, it seems to hang over a neighbor's chimney or just above a particular telephone pole, then it will occupy that spot throughout the night, throughout the year—and, indeed, throughout your lifetime. Let the sharply etched winter constellations surrender to the hazy patterns of summer. Let the Milky Way whirl its gentle sheen into new angles and configurations. Let the various planets display their wares and then drop into the sun's glare to vanish until another season: Polaris remains glued in place. "Constant as the northern star," said Shakespeare's Julius Caesar, longing perhaps to link to it his own ambitions for immortality.

This inertness, then, is what makes the North Star unique. To ask the single stationary object to be *also* the brightest star would be demanding a bit much. That would be a miracle requiring nothing short of a religious explanation.

It comes pretty close as is. For in all of history, as north stars slowly changed with the passing of millennia, we've never had one as bright *and* as close to true north as Polaris.

The North Star sits motionless in the sky not through some talent of its own, but because our planet's axis of rotation points its way. You can demonstrate this principle by performing a pirouette in a room whose walls and ceiling are splattered with polka dots. (Why not? What are you doing this evening that's more interesting?) As you whirl, the dots seem to whiz by in a blur. But the ceiling dot directly above your head

A one-hour time exposure facing due north would show this much apparent motion in the stars. Polaris is the dot in the center. Ursa Minor is directly below Polaris; Ursa Major is at lower right.

On a January evening the Big Dipper lies above the Hudson River near Kingston, New York, the pointer stars aligned vertically beneath Polaris, nearly halfway up the sky.

In Clearwater, Florida, the Dipper takes a dip, only a dubious Dubhe on the horizon to help you find Polaris less than three fists above.

appears fixed in place as you spin, for it alone remains above the axis of your rotation.

Our planet Earth is the eternal dancer. If we stood at the North Pole we'd need only look straight up to see Polaris hanging motionless while all the other stars lazily circled it counterclockwise in the course of nearly twenty-four hours. Observers increasingly far from the North Pole find Polaris standing askew at greater distances from the zenith, until at the equator it sits just above the northern horizon. But even there, where people live on the "sideways" part of the planet, the night's objects wheel around Polaris. Regardless of your location, the North Star seems like the stationary center of a phonograph record or compact disc while all else spins tirelessly.

The height of Polaris equals your latitude. (Don't know your latitude? Neither do most people; check an atlas.) If you live in Miami, latitude 26, the North Star is just 26 degrees high. From New York, latitude 41, Polaris is 41 degrees above the horizon, almost halfway up the sky. At the equator, latitude 0, it hugs the horizon; while at the North Pole, latitude 90, it is, as we've seen, straight up—90 degrees.

If you're lost, you can therefore determine your latitude by simply measuring the height of Polaris. Because a clenched fist held at arm's length subtends an angle of very nearly 10 degrees, you're never at a loss for a

The southern sky above Cape Town, South Africa. How do you find true south without a polestar? There are three aids: (1) Crux, the Southern Cross (top center in the Milky Way); (2) the Large Magellanic Cloud and Small Magellanic Cloud (patches of light above Table Mountain, LMC to the right, SMC to the left); and (3) Canopus, second-brightest star in the heavens (on the other side of the LMC from the SMC).

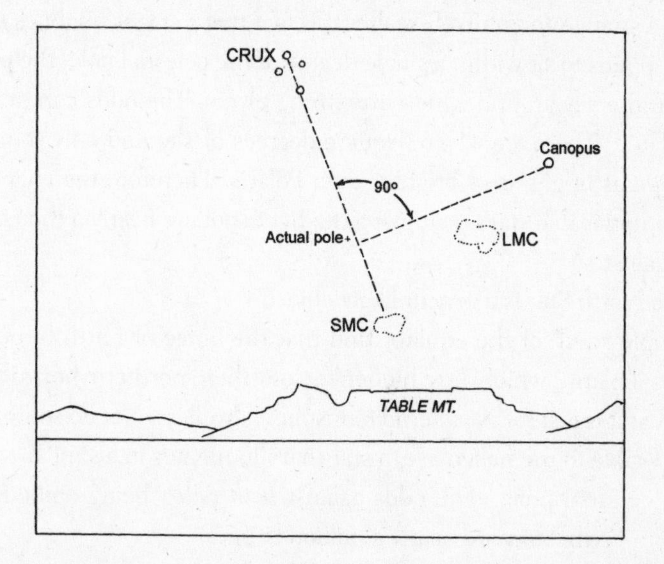

The upright of the Cross points down to the south celestial pole and to the Small Magellanic Cloud on the other side of the pole. A perpendicular from Canopus to the Crux-SMC line fixes the position of the pole. Even with clouds covering one or two of our aids, we can find the pole with practice: a hair less than four extended fists from Canopus, a hair less than three from Crux, and one and a half from the SMC.

Establishing true north with Polaris and a plumb bob

measuring instrument. You can always find your location (distance from the equator) to within a degree or two.

It's a strange and unlikely reality, this fact that a star as bright as Polaris just happens to sit within a single degree of the celestial pole, the precise spot in the sky around which everything pivots. The odds can be easily calculated. There are 41,253 square degrees of sky, and only some fifty stars are as bright as or brighter than Polaris. Therefore the chances of such a noticeable star's occupying the right spot are nearly a thousand to one against.

The North Star is a very unlikely object.

People south of the equator find that the bulge of Earth forever obstructs Polaris, which lies hidden below their northern horizon. The North Star is just for Northern Hemisphere dwellers. Nor do southerners have, visible to the naked eye, a star that illuminates in a similar way the south celestial pole. (The odds against *both* poles' being embodied by medium-bright stars are nearly a million to one.)

When Australians want to locate their pole (around which everything spins *clockwise*) they turn to the Southern Cross. Extending the vertical pole of the Cross by five times its length takes one to the correct spot, even though it is just a blank piece of celestial real estate.

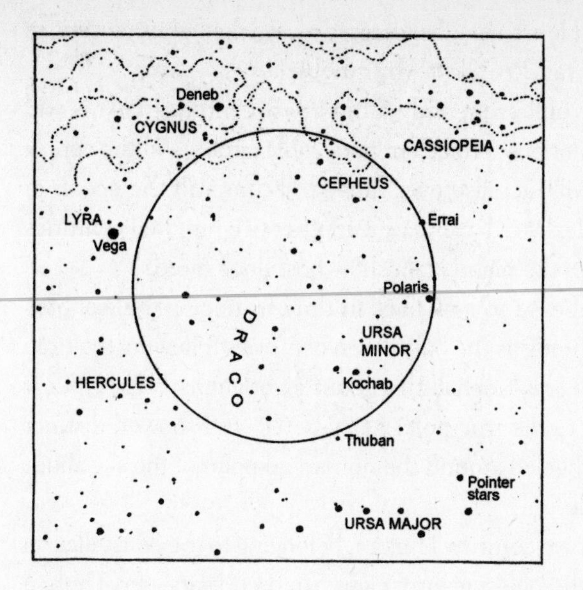

This large circle traces the places in the sky that appear stationary during the twenty-six-thousand-year precession. The distance between Vega and Polaris is a good measure of the diameter of the circle traced out by the motion of Earth's northern pole.

The North Star is, of course, more than merely stationary; it shows us true north. If we drop an imaginary plumb bob straight down from Polaris, it touches the northern horizon to an accuracy of better than a degree. A compass, by contrast, merely follows the magnetic pole, and is badly in error from most locations. From Boston, for instance, compasses point a whopping 16 degrees to the left of north, while instruments on the Pacific coast point nearly as erroneously in the other direction.

The North Star may be constant, but it's not permanent. Over the course of a few human lifetimes it stays fairly stationary, but over longer periods there's a noticeable drift, thanks to a fascinating wobbly motion of our planet called **precession.**

Like a top tilting in a slow circle as its spin winds down, Earth performs a similar wobble every 25,780 years. In that span, our axis points to various places in the sky, tracing out a giant loop of 47-degree diameter.

Of all the stars that happen to lie within a degree of that band, Polaris is the brightest! This means that in the entire twenty-six-millennium precessional wobble, we cannot have a brighter, truer North Star than the present one.

The celestial pole, the precise place to which Earth's axis points, is still slowly marching in the direction of Polaris. Now within 1 degree of the pole, Polaris will stand less than a half degree away when at its closest, in the year 2106. (Some say 2012, but why quibble?)

Then it will slowly pull away, and within a few centuries Polaris will start to exhibit small motions—trace out noticeable circles—in the course of the night. These will increase over time, and thus will the epoch of Polaris slowly conclude. We'll have to wait another two hundred centuries before Polaris becomes the nightly talk show host once more.

It's fascinating, and easy, to look back in time by tracing the wobble's path in reverse. As recently as the half-dozen centuries following the time of Christ, the polestar was Kochab. It's about as bright as Polaris, but it never really got close to the true pole. At its best it still showed distinct nightly motion as it wheeled around the unmarked point of the sky about which the stars revolve.

Going back further we come to Thuban, belonging to the constellation Draco the Dragon—the polestar forty-eight hundred years ago. Thuban did indeed sit right smack at the celestial pole, a superb polestar, though only one-fourth as bright as Polaris. Moses, looking up into the desert sky, would have seen it standing perfectly motionless. But forty years of wandering may have meant he wasn't paying too much attention to directions.

With Thuban we encounter a mystery that still hasn't been answered to everyone's satisfaction. The puzzle surrounds the Great Pyramid, the most massive and precisely aligned structure in the world, whose construction also occurred during Thuban's reign. The pyramid's main passageway is directed to a point in the sky that Thuban would have occupied about six centuries before or after its nearest advance to true north.

If we assume the pyramid builders aimed this passageway at Thuban, their North Star, it permits us to date the structure as having been built in either 3500 B.C. or 2200 B.C. The problem: Scholars are certain that neither date is correct. If so, we're left with the bewildering conclusion that the pyramid's architects deliberately pointed the passageway to a blank, meaningless spot of sky, a place that Thuban would someday occupy in their future, or where it had previously dwelled. Why would they do this?

Sloppiness or error seems out of the question. The sides of the pyramid are constructed in line with the cardinal directions to an accuracy of better than a tenth of a degree. Why should the descending passageway point 3 degrees away from the North Star of that era?

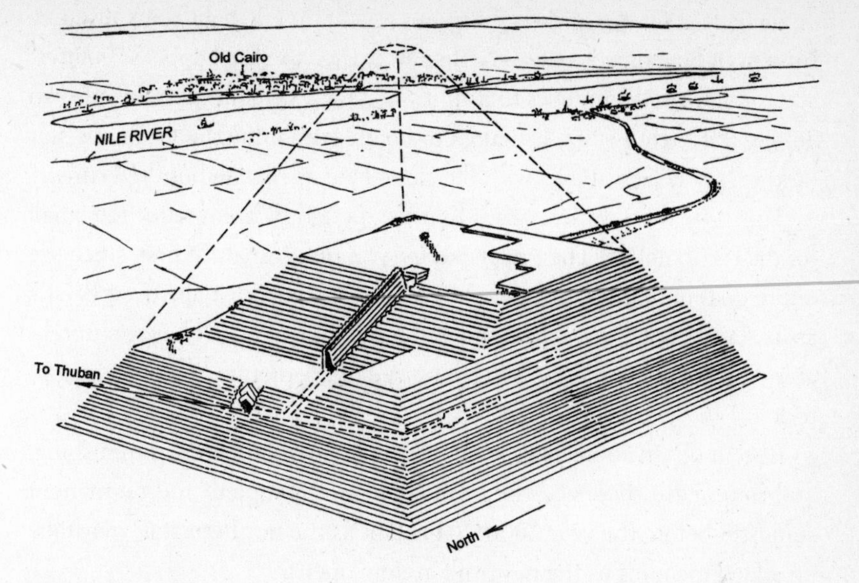

The Great Pyramid at Giza under construction. The pyramid is precisely aligned to true north. The descending passage (shown with dashed lines) supposedly pointed toward—but not precisely at—Thuban, the North Star of that era.

Debate has raged among archeoastronomers and historians for more than half a century. One of the more recent theories posits that the passageway never did point to the builders' North Star at all. Another view is that there was nothing deliberately esoteric about the layout. Prosaic as it may seem, the angle of the passageway happens to equal the slant one gets by laying several blocks across for every one up. In short, the builders may have been concerned only about employing the easiest method to build a ramp!

Seven thousand years before the pyramids, the brilliant stars Vega and Deneb took turns being the North Star, as they will again some twelve thousand years in our future. While these are, by far, the brightest stars to venture near the celestial pole, neither will ever be particularly close to it, or appear stationary in the night. And between now—the age of Polaris—and those distant millennia, no star will dominate. This, then, is a truly memorable era we're in, this unlikely Age of Polaris.

When we examine the star itself through the mind of science, we learn that the distant orb to which Earth's axis happens by chance to point is remarkable in its own right.

First off, it's not an ordinary star, but a giant. At 600 light-years' distance from us, it lies some six times farther away than the Big Dipper's "pointer" stars, which guide our eyes to it. If it's that far away and yet holds its own among the night's stars, it figures to be an extraordinarily luminous star (and in fact Polaris shines with the same light as sixteen hundred suns).

This brightness varies over a four-day period by an amount too small for the eye to notice. The tiny flickering was unfortunate at first, since the entire system of measuring the brightness of stars originally used Polaris as its standard. Polaris *defined* magnitude 2. But you can't have an unreliable standard star, and the system moved on, demoting Polaris to magnitude 2.1 in the process.

Recent observations have shown that the fluctuations are diminishing, and some have theorized they will come to a complete and permanent standstill before the year 2000. "Constant as the northern star" may take on added meaning as the new millennium opens.

Since every celestial body in the cosmos rotates, we didn't need to be born on Earth to see the whirling of stars around a celestial pole. It's a universal experience. No other known world has an axis pointing to Polaris, of course, nor is the rate of sky motion similar to ours on any other planet except Mars, where the stars take an extra half hour to circle the sky each day.

The greatest sluggishness in the solar system is found on Venus, where it takes not twenty-four hours, but 225 days for any star to cross the sky, set, and then reappear. Conversely, the speediest sky would be seen above Jupiter, where the stars cycle the heavens in under ten hours. But it's academic: Both worlds are overcast.

For a truly snappy spin, you can look beyond the solar system to the tiny collapsed pulsar in the Crab Nebula. There the night's stars traverse the sky in a dizzying blur every thirty-third of a second. The mad streaking of the stars around their north celestial pole would appear to fashion a series of solid concentric rings. Other ultradense crystalline stars spin so frantically that everything in their sky rises, crosses the heavens, sets, and rises anew *hundreds of times* each second.

But Polaris symbolizes the antithesis of such jittery motion. When observing it telescopically, one has the unique pleasure of being able to gaze with leisure. While all other celestial targets quickly drift from view because of Earth's rotation, observing Polaris offers the singular boon of admiring a specimen that seems pinned to the table.

This is the one celestial target for which no tracking equipment is needed, and this lack of motion suggests several lazy projects for the slightly ambitious nature buff. The easiest is to set up a camera on a tripod and open the shutter for a few hours. The whirling, counterclockwise motion of the stars orbiting Polaris makes for a dramatic photo. One caution: The picture must be taken far from the lights of a big city or the film will fog (overexpose) in a few minutes.

Readers who have small telescopes with lenses or mirrors at least three inches in diameter are able to detect a small companion star to Polaris, a wonderful sight. Studies of the light of Polaris indicate a third, unseen star as well, so that the North Star is actually a triplet.

Then there is another, more restful possibility. Point your instrument at Polaris and leave it there. Take a twenty-, thirty-, even forty-year sabbatical. Return filled with a lifetime's joys, regrets, and accumulated wisdom—and the constant star will still be waiting for you.

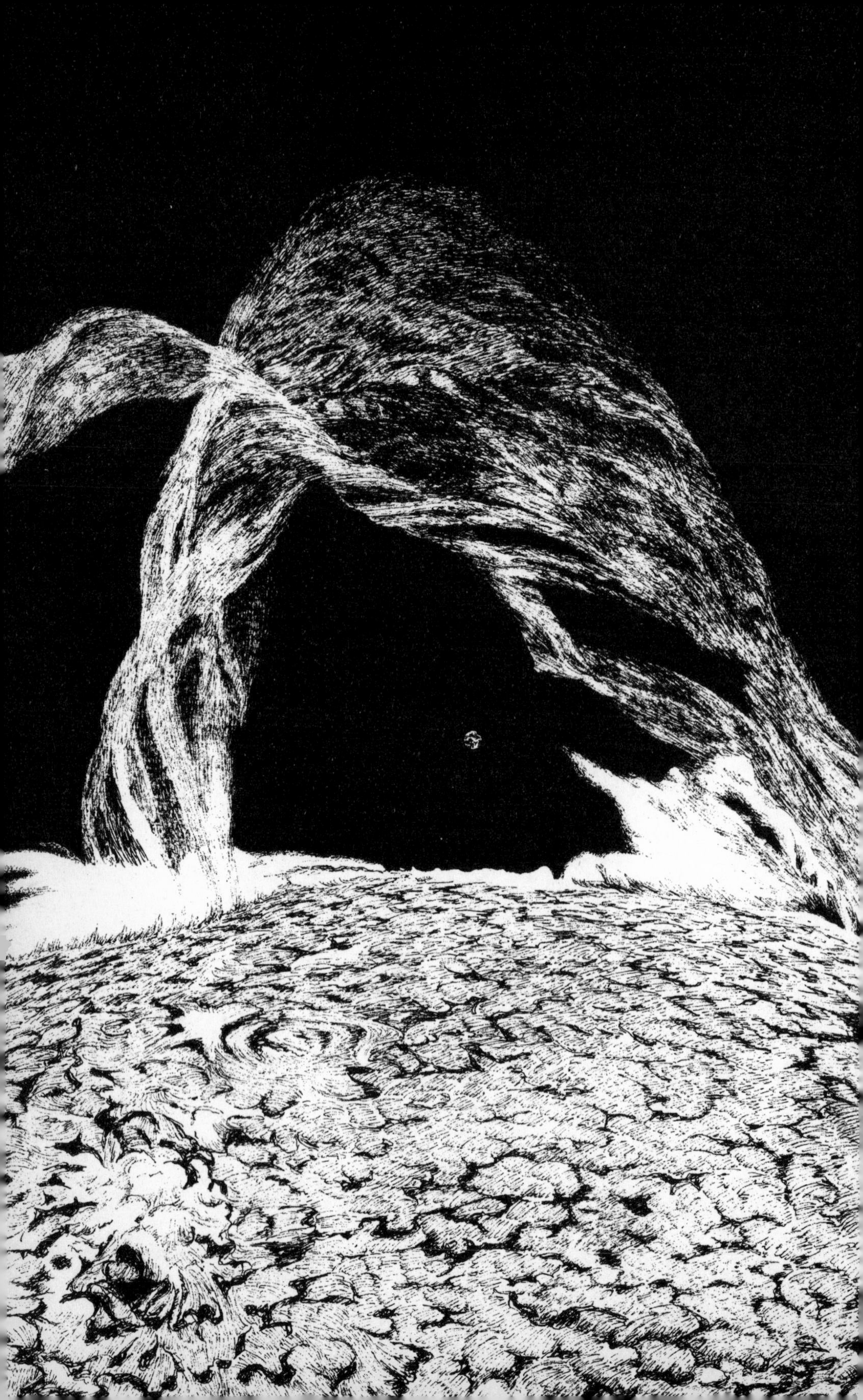

Spring

An astounding solar prominence observed in ultraviolet light (see the chapter "The Favorite Color of the Universe"). For scale, Earth hovers under the arch.

Walking in from Arcturus

Spring's brightest star reminds me of a furor a few years ago in Woodstock, New York. In a town where nothing is surprising, a famous "psychic" drew crowds by claiming he was inhabited by a being from Arcturus. No big deal, he said; many people's brains are occupied by Arcturians. Such uninvited neural advisers were called walk-ins, and this particular walk-in announced an ambitious lecture series throughout the United States. Needless to say, the audience paid an entrance fee—in terrestrial cash.

Such astroscams are easy to debunk. The "psychic," like most people, was apparently unaware that only a few dozen stars have popular names in common use. A million other cataloged stars have designations like HD212710. If aliens *were* to arrive, it's suspiciously unlikely they'd just happen to come from one of the few stars with a catchy name like Arcturus.

But if I were a walk-in I'd be proud to hail from there. And not just because it's the brightest star seen high overhead from our part of the world. Arcturus is a unique, amazing object, the only celestial body to open a world's fair. And the only major star that will soon . . . disappear!

Even beginners spot it easily: the Big Dipper's handle curves in its direction. "Arc to Arcturus" has been counseled to young stargazers almost as often as "Don't put your fingers on the lens." (See figure, page 108.)

On early spring evenings, when Orion's army of jewels crumbles into the west—and while the high northern sky offers the year's best view of the Big Dipper—the east belongs to solitary Arcturus. Its pumpkin-colored rays emanate from a sphere so large that 25 billion Earths could fit inside.

This staggering size, together with a fairly nearby location, combine to make Arcturus the second-brightest star visible from our region of the world. In all the heavens it's bested only by the Dog Star, Sirius, and the two jewels seen exclusively from tropical and southern climes: Canopus and Alpha Centauri.

Still, one suspects that the name Arcturus, easily pronounced and readily remembered, plays a big part in its stardom. While most star names come to us from the ancient Arabic and are far easier to find than to utter (when was the last time anyone dropped Zubeneschamali in conversation?), Arcturus is of Latin origin, easier on Western ears. Its name means "guardian of the bear" (as if bruins needed bodyguards!)—an allusion to its proximity to that fellow citizen of the spring sky, Ursa Major.

We may as well digress about stellar nomenclature, and go back to the dismaying fact that catchy names like Orion and Arcturus belong to members of a very small minority party. Most of the sky is crowded with faint patterns offering few associative possibilities, and sporting wildly esoteric names. In this corpus of incoherency Arcturus stands happily apart.

Unless one abandons sobriety altogether, no amount of imaginative effort could construct a fox out of the dim stars of the constellation Vulpecula. Octans the Octant, one of many patterns recklessly defined by mariners of the sixteenth and seventeenth centuries, is so faint it's difficult to find at all. Aquarius is a haphazard connect-the-dots that ambles senselessly. Capricornus the Sea Goat (whatever *that* is) boasts the outline of New York State. Small wonder that learning most constellations is harder than deciphering an insurance policy; the few easy, rational patterns and Arcturus-type names seem welcome harbors in a sea of hieroglyphics.

It might be nice to start all over with newer, more contemporary labels. But our modern objects, animals, and job titles seem almost as preposterous as Ophiuchus the Serpent Bearer, seen every summer. While such reptile portage is no longer a profession in demand (the serpent bearers' union a mere shadow of its former strength), today's more typical careers seem just as inappropriate for the heavens. The Chiropractor? The Bail Bondsman? The Insurance Salesman?

Nothing works. And that's probably why, on starlit nights, we'll just go on teaching our children to look upward for the ancient constellations even when they sound ridiculous or are essentially blank areas of sky.

Those of us living in the Northern Hemisphere are fortunate, at least, to see constellations named mostly in classical times. Although many are obscure or dim, they've at least had the benefit of screen testing by a hundred human generations. The same cannot be said of Southern Hemisphere patterns. Spied for the first time by Renaissance explorers sailing under unfamiliar skies, the southern constellations resulted from a binge of star naming in honor of the unfamiliar species of birds and animals they saw and the instruments that got them there. Thus we have Tucana the Toucan and Volans the Flying Fish, along with a warehouse full of antique instruments like Sextans the Sextant and Antlia the Air Pump.

The ancient mariners apparently didn't foresee that their love of gadgets would drive future sky watchers bonkers, as we now try to envision a "Square Rule" in the random stars of Norma. It would be like coining new constellations today called Cordless Phonium or Satellite Dishium. Contemporary technology is a poor template for celestial immortalization: It not only fails to stand the test of time but suggests, to future generations, the deification of doodads.

Arcturus, on the other hand, is part of a constellation formed from easily traceable stars that do look a bit like a human figure, if not the specific *herdsman* it's supposed to be. It's still not all peaches and cream: Boötes, its constellation, has thrown beginners for a pronunciational loop for thousands of years. Once we get the hang of splitting it into three syllables to correctly say boo-OH-teez, it's not so bad. Better yet is to forget the humdrum stars surrounding brilliant Arcturus like sycophants—in short, ignore the constellation—and simply focus on the beautiful yellow-orange giant itself. This is easy to do, since it dominates the entire eastern sky as darkness falls from March to May, and then stands high overhead throughout the early summer.

Arcturus is an immense orange sun. Ordinary stars of this color are hopelessly invisible because they have a low brightness per square foot.

Arcturus does, too, but compensates for this by being so big, boasting such an immense radiating surface that it puts out as much light as a thousand normal stars of the same golden color.

Its brightness and fiery tint form a striking contrast against a blue sky. This may explain why Arcturus was selected for an all-but-forgotten distinction: It was the very first star ever seen in broad daylight through a telescope, an event occurring nearly four centuries ago. These spring nights it's the starlight, star bright, first star you'll see emerge in the sky, a memorable apparition through binoculars against the deepening azure of mid-twilight. At such times it's easy to imagine its scorching nuclear intensity, a solitary ruddy ember emitting more than a hundred times the brilliance of our own sun.

Could you, by raising your palms toward the eastern sky, feel some of that heat? Not likely. According to astronomer Robert Burnham, the warmth we get from Arcturus is equal to a single candle located five miles away.

But that energy was put to good use back in 1933 for the opening of the famous Chicago World's Fair. Arcturus' light was focused through a telescope onto a photocell (a newfangled device back then) that tripped the lights to start the festivities. Arcturus was chosen because it was thought to lie 40 light-years away, so that the same light that had left the star forty years earlier—when a previous Chicago fair was closing—would open the new one.

A wonderful and romantic idea, a salute to continuity rarely expressed in our fast-paced modern era. It doesn't even seem to matter so much that they were wrong about Arcturus' distance. It's now known to lie 10 percent nearer, 36 light-years from us, so that the rays opening that most famous of all world's fairs were just bits of everyday Arcturian light after all.

But Arcturus will not stay put. Restlessly marching to a different drummer, it follows a peculiar path around our galaxy's center, avoiding the galactic plane in which our own sun and most other stars orbit. Alone among the night's bright stars, it springs upward out of the Milky Way's spiral flatness, only to dive through it 100 million years later, going the other way. Instead of traveling along together with us like an adjacent horse on a carousel, Arcturus is plunging down at us from above.

Arcturus' orbit (solid arrow) is perpendicular to the plane of the Galaxy and the sun's orbit (dashed arrow). Both are 30,000 light-years from the galactic center.

This causes rapid changes in its brightness and location. Just five hundred thousand years ago—a hiccup in time—Arcturus was invisible. It's been steadily approaching us ever since, and is 1,000 miles nearer since you started reading this chapter. Skimming through our neighborhood, it's now almost at its closest and brightest point. Then, half a million years from now, before our Earth and sun have revolved even a two-hundredth of the way around the Galaxy's center, Arcturus will have faded into oblivion forever.

When we next return to this part of our orbit, Arcturus and its enterprising walk-ins will be somewhere else. So greet it warmly and pay it a salute during this one cycle in time when our paths cross.

For we will never meet again.

Other Universes

Bored with the entire universe?

Consider others. Cloudy nights, when the cosmos is cloaked, are the perfect times to do one's exploring inside the mind, by probing the prospect of parallel dimensions. Sit back. We're going on a breathtaking tour that begins and ends at the fringes of reality.

In earlier, simpler times, *universe* meant "everything," and that was the end of the discussion. But the word, like everything else, has evolved. Now it often means "everything currently and *potentially* observable," so that if a domain, for various reasons, must forever lie beyond our senses or instruments, it may be regarded as a separate universe altogether.

That's a good working definition, allowing us merrily to explore parallel, tachyon, temporal, and other hypothetical universes without needing a shred of evidence that they actually exist.

Let's start by shedding light where light can never reach: the tachyon realm. This is the land where *everything* travels faster than light, preventing its images from interacting with us.

Any exploration of such a superluminal, or faster-than-light, dimension must begin with Einstein's special theory of relativity, formulated in 1905. Relativity holds that nothing that weighs anything can ever attain the speed of light—186,282.4 miles per second. Nearly a century's experimentation has confirmed its validity. As in one of Alice's strange adventures, your mass increases as you accelerate. You'd outweigh the entire universe—a dieter's worst nightmare—before you could attain light speed, even if you began your odyssey lighter than a fragment of incense smoke. No amount of energy could shove your newly obese body farther, let alone beyond—to superluminal velocities.

But relativity contains a curious loophole, as if lobbyists from another dimension managed to crash the rule-making council: It applies only to voyagers who *start off* at slower-than-light speeds. At first this sounds like no help at all, since everything in the cosmos falls in this category. But here's the catch: Nothing forbids the existence of a separate reality where everything has *always* traveled faster than light. Such superluminal objects are called **tachyons.**

When and if the Big Bang spawned the current universe, a separate tachyon dimension could have originated simultaneously—or at least it's permissible mathematically. Since it takes as much energy to decelerate as to speed up, residents of tachyon worlds find, just as we do, that the speed of light is an insuperable barrier. Neither of us can cross it. And while we cannot *accelerate* beyond it, they're on the other side. They have no way to *slow down* to light speed. We can never meet. We cannot even see each other.

Do tachyon particles, tachyon planets, or tachyon people exist? Nobody knows. But it wouldn't be advisable to get involved in a tachyon relationship: too fleeting. We couldn't even wave to each other. However, it might just be possible to detect their influence at the subatomic level. Experiments attempting this have proven negative to date, and most physicists regard tachyons the way the rest of us view the stories in supermarket tabloids. Still, tachyons do have their scientific supporters, and the truth is, we do not know.

If life in such a fast lane makes you jittery, then travel a different route—to the quantum foam lurking in the realm of the tiny.

Quantum foam is not a birth control product, though it is related to conception. It involves the birth of infant universes.

Permissible in the arcane kingdom of speculative physics, this concept has no more observational support than tachyons. Still, it's theoretically possible.

Most physicists now believe that on the very small scale, a million times tinier than a trillionth of an atomic nucleus, space itself is a sudsy turbulent froth, like a recklessly oversoaped bubble bath. Stephen Hawking maintained that bubbles of spacetime may continually break away from this foamy "surface," connected only by the thinnest of wormholes—narrow umbilical cords far too slender to allow even subatomic particles to migrate across.

Instantaneously upon creation, each of these tiny bubbles experiences its own Big Bang, swiftly growing into a full, independent universe connected to ours only by those uselessly narrow threads. Then, like the wildly multiplying brooms in Disney's "Sorcerer's Apprentice," each new universe vigorously forms its own offshoots. Branch after branch, bubble upon bubble, endless universes arise like dough expanding in some hyper-yeasted cake.

Nor does this happen infrequently. Every fragment of existence, even your junk mail, continually spawns new and complete universes. The exact number? According to the theory, each sugar-cube-sized piece of our own universe gives birth to a hundred trillion times more than a trillion trillion trillion trillion trillion trillion trillion trillion trillion trillion trillion new universes each second.

Just in the combustion cylinder of a lawnmower engine, there's a continual creation of 100,000,000,000,000,000,000,000,000,000,000,-000,000,000,000,000,000,000,000,000,000,000,000,000,000,000,-

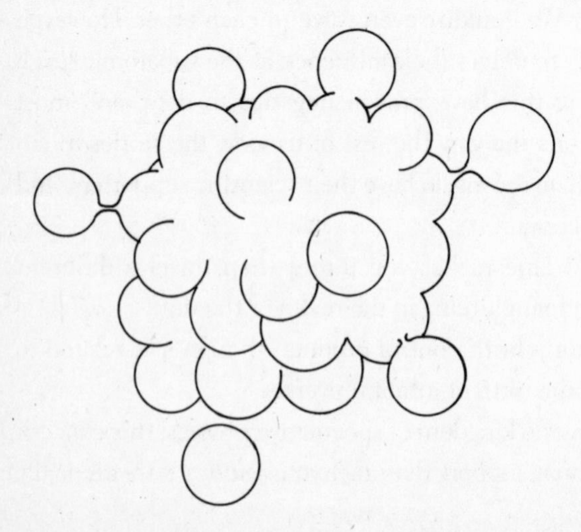

On the extremely small scale, reality may be sudsy. Each breakaway bubble might become a complete, independent universe.

000,000,000,000,000,000,000,000,000,000,000,000,000,000,000,000,-000,000,000,000,000,000,000,000 entire universes every second, each of which contains all the galaxies, planets, and potential life that ours does. Finally, something greater than our national debt.

Unfortunately, all this cosmic procreation must remain forever unobservable, since not even the wavelength of the most energetic light could make the journey across those constricted wormholes to let us see how the other folks live.

If one suffered from the "grass is always greener" syndrome, one could be fairly miserable contemplating the hidden universes that, at minimum, sprout from cosmologists' minds.

But in such lavish alternative realities, is there at least a common *temporal* thread? In short, is time the same everywhere? Can it flow in different directions in other dimensions? We've looked at two possible spatial alternatives to our own universe. Can we find others separated from ours by *time*?

It's an important issue, since even exploring the workings of our own little corner of the cosmos forces us to take time into account. Certainly, cause and effect demands a direction or current in the passage of events. Is that the case everywhere and always? The question itself reeks of paradox. But the answer determines the fate of the universe—or at least our judgments about it.

There are many areas in which time does *not* seem to have a directionality. The laws of physics are almost entirely time-symmetrical, and that includes quantum mechanics, relativity, and Newton's laws. In these, time doesn't matter. It can run backward or forward and produce the same results. For example, a movie of planets orbiting the sun, shown in reverse, would not appear incorrect. You couldn't even tell whether the film was being shown backward: The counterclockwise revolutions as seen from north of the solar system look like a time-reversed video of orbital motions viewed from the south, the other side. Gravitational interactions are valid either way, a two-way street.

Such universal laws contradict our innate sensibilities as well as everyday experience, where most events appear to unfold in just one direction. Reverse a film of an auto accident or an erupting volcano and you'd instantly know that things were screwy. Science fiction has often played

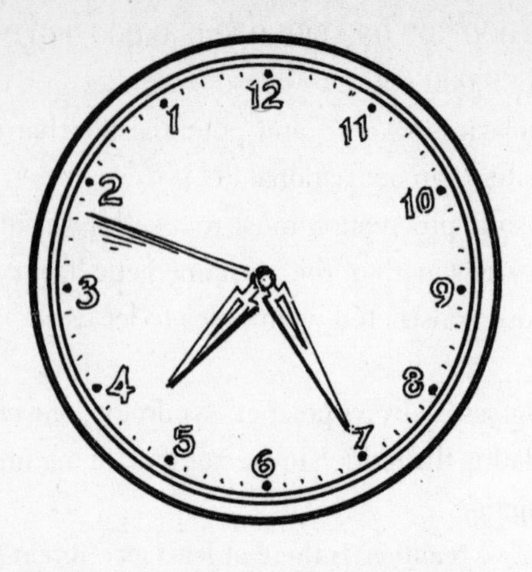

Can time flow backward?

with the idea of topsy-turvy time, and Hawking once declared that when and if the universe starts contracting, time will flow the other way. (But later he changed his mind, as if to demonstrate the process.)

It's hard to visualize reversed time. Effect could never precede cause—by definition. In the case of a midair collision, could we ever accept that pieces of metallic rubbish lying on the ground might jump into the sky, assemble themselves seamlessly, and then fly away backward as two perfectly functioning airplanes? Some have tried to get around the problem by suggesting that our own mental processes would run backward as well, so we'd never perceive anything amiss.

It's also possible that time behaves uniquely on different planes. Perhaps levels of reality where time is important coexist with other planes in which it simply does not exist at all.

There are serious problems with any dimension operating in a time frame different from our own. Even tachyons produce instant causality problems, and as for *separate* temporal realms—it's hard to imagine how we could ever interact at all. Of course, that's why they'd be considered a sequestered universe to begin with.

If you don't have time for all this temporal speculation, we can quickly shift into yet another kingdom by hurling ourselves into a black hole, the ultimate threat in a lover's moment of pique. (Black holes have their own chapter, so we'll confine ourselves at this point to their effect on other dimensions.)

Space and time warp severely in a black hole's immediate environment, yet still function in familiar (if distorted) ways. To an onlooker watching you leap toward that superdense object, you'd appear to slow down as you approached the **event horizon,** the boundary where space and time curve back unto themselves like a breaking wave. As you reached it, reality would split neatly in two. There'd be you as others see you, and you as you see yourself. This may sound a bit like the way things already are in everyday life, but the schizophrenia runs deeper when near such a collapsed star.

You see yourself falling ever faster toward the black hole's center, yanked so violently that your leading limbs are accelerated at a much faster rate than your trailing limbs. In plain English: You're ripped to shreds. If, somehow, your spirit or consciousness could endure such mistreatment, you'd quickly arrive at the singularity at the black hole's center. A **singularity** is a place where an entire star weighing 2 million planet Earths has been crushed down to less than the size of an atom.

Actually, a singularity is really much, *much* smaller than an atom. An atom would be the Grand Canyon compared to a singularity. A singularity occupies no space at all. And you—hapless, impulsive you—now occupy no volume either. Talk about feeling small.

What is life like in the zero dimension? Physicists have an unhesitating answer. They say: We have no idea. For the very laws of physics break down, and nobody has suggested new ones for this Land of Nothing. Our universe leaves off. Something else, presumably, begins. It would have to be a "something" that occupies no space in our own dimension.

It was suggested by Einstein and the mathematician Nathan Rosen that a tunnel, a wormhole, could provide a conduit for the black hole and its contents (in this case, you) to reemerge in another place or time. Such a geyserlike entrance back into our own universe has been termed a **white hole.** As recently as the late sixties, some theoreticians were speculating that quasars, now known to be the explosive hearts of young galaxies, were actually white holes where material erupted into our universe after having left other universes or other *times* of our own universe, the consequence of having fallen into a black hole somewhere else.

But, as has been noted, quasars have been explained, and they are definitely not alien shuttle flights through other kingdoms. If a black hole is a bridge to another time—and most theoreticians now dismiss this as unlikely—we do not yet see its material reemerging here. It's there on the other side, whatever and wherever (or whenever) that might be.

Meanwhile, let's not forget your better half, that other image of yourself that slowed down as you were entering the event horizon of the black hole. People watching safely from afar, perhaps to see if they wanted to boldly go where no prudent traveler had gone before, might conclude that you'd had a change of mind and come to your senses. For they'd see you approach the black hole only to grind to a halt and freeze at the boundary, rather than proceeding further. And there you would remain, motionless for lifetimes, your image slowly fading like an old photograph. No, neither part of your split personality seems to be having more fun than the other.

It is time, then, to try out another dimension, one with more potential for entertainment: a duplicate Earth! Questions about its existence continue to arise in astronomy classes, largely inspired by a 1969 science fiction film in which our planet was portrayed as having a twin.

The usual scenario involves a second world occupying our orbit, but lying on the far side of the sun. This way, so the reasoning goes, it would always be hidden from view. Supposedly it is peopled with our reversed analogues, who possess strengths where we have weaknesses. Life is different there: At night you remove your socks neatly instead of tossing them, that sort of thing.

The frequency of questions about this make-believe world shows that it's a popular idea, but it's easy to disprove any possibility of an unseen planet in the inner solar system. Among many other reasons, it would betray itself by gravitational influences on other planets, comets, and the half-dozen spacecraft that have whipped past that region.

Finally there's the oldest idea of all, that our familiar universe of planets, stars, and fast-food restaurants is All There Is. Einstein made it easier, settling the old confined-versus-infinite debate by characterizing our own dimension as limited but boundless.

This means that the universe contains a specific, fixed amount of material and energy, but in a matrix of curved space that forces everything, and any traveler, to follow a looping path. No matter the route, one never encounters an end, boundary, or barrier. But no infinity, either. It's finite but unbounded. Given enough time, patience, and salary you could count every star. Travel from Sheboygan for eight dozen billion years in an apparent straight line, and you still end up there.

In reality the equations are not that simple, and your giant arc would not likely culminate at your starting point. But it wouldn't matter. One lap around the universe would take so long that the entire cosmos would evolve in the interim. Traveling near light speed and enjoying the age-extending benefits of relativistic time dilation, your body could last long enough to permit such an enlightening circuit, but you'd be carried into an epoch so utterly advanced that the planet you left would be unrecognizable upon your return, if it still existed at all. (See "The Once and Future Past," page 242.) It might require, to use real numbers, 100 billion years to circle the universe at an eyelash below light speed. In that interval, even if less than a human lifespan simultaneously passed in your own reference frame, our sun could complete a half-dozen lifetimes from birth to frozen collapse. At the end of your trip, the solar system's death would lie 90 billion years in the forgotten past.

Our vast universe of galaxy clusters and quasars, the realm of the known and the explorable, could indeed be the Whole Shot. Or it could all merely be an electron in a toothpick in some back-street diner in an incredibly larger realm.

Either way, one of its most curious designs may be the mechanism contemplating the issue. Conveniently constructed of water, the most common compound in the cosmos, it is the human brain—a universe of its own.

The Favorite Color of the Universe

With winter's gray surrendering to the diversity of spring, and with daylight increasing by nearly four minutes a day, a spectacular dawning of color now surrounds us. But what of the night sky?

Is it merely, as many believe, a simple case of light dots on a dark background? Or, conversely, does space really resemble the breathtaking spectrum of colors found in magazines and textbook photographs? In both cases the answer is no. The true appearance of planets, stars, and the universe is very different from what most people imagine. Nor, when it comes to color, is there equality. The universe shows a strong preference for one particular hue, which recurs like a musical rondo throughout time and space.

On any March evening, peer upward to the conspicuous three-stars-in-a-row of Orion's belt. They're our guide to brilliant Betelgeuse just above it, and even more dazzling Rigel an equal distance below. No color in the sky? Just glance at the lovely contrasting tints of these two jewels. Betelgeuse (as we saw on pages 34–35) is as golden as a marigold, while Rigel, like most of Orion's distant suns, shines blue-white. Binoculars bring out stellar colors nicely, but even the naked eye reveals a spectrum of starry pastels. For the night's stars come in red, orange, yellow, blue, violet, brown, and even black. All the colors except green.

But even the existence of green stars is an open question. The confusion surrounding star colors is exemplified by the famous red giant star Antares, which marks the heart of Scorpius the Scorpion, rising these mornings just ahead of the dawn. Telescopes reveal it to be a binary star, two suns orbiting each other. While the brighter of the two is clearly reddish, many texts list the companion as green. But others maintain that the eye simply reacts to the brighter star's rosy tint by imposing the complementary color on a nearby white surface, just as staring at a red traffic light produces a hallucinatory green blob when you look away. They point out that when our moon eclipses the brighter member of the pair, the supposed greenness of its companion vanishes. Such disputes underscore the subjectivity of color perception. And since there is no ultimate authority, no supreme court of color arbitration, reference books continue to contradict each other about the complexion of many stars—a sort of ongoing hue and cry.

In some cases, the insertion of a telescope between beholder and the beheld only adds to the problem. A case in point is Cor Caroli, the brightest star of the constellation Canes Venatici, the Hunting Dogs. Trying to picture the traditional "two hounds" in this tiny star group is a challenging enough exercise in creative abstraction, but with any telescope one can add an additional puzzle: Cor Caroli, under low power, quickly splits into a gorgeous, colorful double star. The issue, again: What color are the components?

The brighter of the two is clearly blue-white. The companion, however, is listed in various respected texts as rose, orange, amber, yellow, white, blue, and violet. Which of those venerable authors had it right? Over the years, I've invited hundreds of people who were visiting my observatory in upstate New York to peek into the eyepiece and cast a vote. The result? An even split. Personally, I've always seen it as pastel violet, but remain outvoted by a wide plurality, probably by people who do not share my lifelong obsession with purple.

There may not be green stars, but that (or at least blue-green) is a popular color for planets. Earth, Neptune, and Uranus sport that tint, the latter appearing so vividly emerald through a small telescope that it is the easiest way to spot Uranus in a crowded star field. But other planets continue to cook up a cornucopia of color confusion.

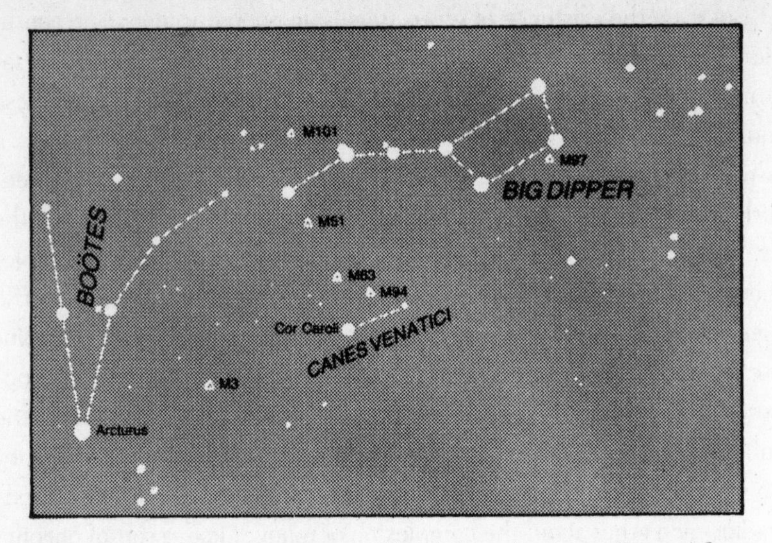

Color contrasts: The Dipper stars are white, Arcturus is orange-yellow, and Cor Caroli is purple. Maybe.

Mars, whose classical association with war doubtless arose when the superstitious Babylonians linked its reddish color to blood or fire, actually has a chocolate-brown surface according to photos taken by the *Viking* spacecraft that landed there in 1976. Had the ancients known the truth, Mars might now be linked with Valentine's Day instead of warfare.

Another famous color-that-isn't is Jupiter's renowned centuries-old storm, the "Red Spot." Thanks to computer-enhanced images (reproduced far more widely than the much duller pinkish-gray of reality) even some astronomers mistakenly assume that the vivid hues of Jupiter's stormy cloud surfaces are real. Textbooks and magazines routinely display photos of "Jupiter" without mentioning that the image has been falsely enhanced. Indeed, the touched-up version has effectively become the reality; the true rendition has scarcely appeared since the 1970's.

As for Venus, until the radar-mapping *Magellan* spacecraft arrived in 1991, the clearest pictures of that hellishly broiling world came from the *Mariner 10* mission in 1973. But even from the probe's nearby vantage point, the planet appeared as a featureless disk except in ultraviolet and infrared light. Ever since, one particular ultraviolet photograph has been consistently published. Because ultraviolet is invisible, there simply is no "true" color. That renowned stock photograph has been reproduced in yellow, blue, you name it, providing millions of readers with an array of random and contradictory Venusian tints. Venus, like Jupiter and Saturn, actually appears as a warm white to the naked eye.

Speaking of ultraviolet, if it were visible to our eyes it would be one of the most common colors in the universe, since that realm of powerful energy lying beyond the blue horizon permeates the galaxies. Add other invisible aspects of the spectrum—the X rays, gamma rays, infrared, microwaves, and all—and you have most of the energy that exists, period. We see none of it. Chalk that up to the limitations of our sun. Humans are attuned only to those wavelengths that the sun emits most strongly. Our sun is a very weak emitter of gamma rays, and sure enough, we're blind to them. This can't be coincidence. Obviously our vision evolved to operate in daylight. We scan the universe with the sun's eyes.

While we're thinking about the sun, it really isn't the yellow star popularly imagined. Its flaxen hue is an artifact of our own atmosphere. As its light beams penetrate air, the shorter (blue) wavelengths are bounced around by the atmosphere, giving us our blue sky. Once this blue has been subtracted from the sun's light, it leaves the sun itself with a warmer, more yellow appearance than it actually possesses. When it's low in the sky and its light is forced to pass through many miles of extra atmosphere, the effect becomes so strong that the sun appears deeply yellow or even orange.

Its true color is revealed by snow, which does a good job of reflecting all wavelengths of light equally. The color of snow on a cloudy day (when the sky's blueness isn't mirrored by the ground) reveals the actual color of the sun. The sun's factual whiteness also becomes obvious when viewed from above the lower, thicker layers of our atmosphere by astronauts in space and high-altitude balloonists. They report that the sun is a pure-white star.

But even this whiteness is an illusion! In reality the sun emits a peacock's display of blues, reds, greens, and violets—the colors projected by a prism. We should remember that a prism or rainbow does not somehow manufacture colors, but rather displays the wavelengths actually vented by the sun. But while a prism or spectroscope obligingly lays them side by side for our inspection, the naked eye swallows them en masse, blending them into what subjectively seems a pearly whiteness. It's similar to the white that magically appears when colored stage lights are all turned on. So we're left with wheels within wheels of chromatic deception: The sun appears yellowish but is really white, even though white is a color falsely perceived by our eyes when combining the true solar colors. What a swindle!

It is not an evolutionary goof that we cannot see the big dose of ultraviolet that the sun emits. To our benefit, our atmosphere blocks more than 99 percent of these dangerous rays. Our batlike insensitivity to them results from their relative scarcity here at the surface.

Another common color myth involves the aurora, or northern lights. Though it is popularly thought to display a rainbow's profusion of color, the truth is again paler than fiction: An aurora is usually just an ashen green. If another tint appears, it will normally be a rich ruby-red, and that's about it.

Even the night sky, seen thousands of times by everyone, is widely misjudged. Most people picture it as black rather than its actual apparent color of dark blue-gray.

And the moon? Despite many romantic but atmospherically induced orange and yellow moonrises, it's one of the grayest objects in the universe. Astronauts who orbited or landed merely wasted color film on its monochromatic surface. As for the notion that it can ever appear blue, even "once in a blue moon"—forget it.

We have seen that the solar system abounds in chromatic illusion. Now let's take a giant step outward, to the universe beyond the planets.

The surprise begins the moment a telescope is pointed at the heavens. Expecting a colorful, swirling nebula like the photographs in magazines, the young astronomer is puzzled and disappointed by the nearly unvarying monochrome. Nothing's wrong with either eyes or instruments. Because humans cannot perceive color under low-light conditions, all of the universe's countless galaxies and star clusters appear gray, even through the largest telescopes. No one has ever seen deep-space objects in the tints displayed on photographs. Film reveals the color of dim subjects because it can do what the eye cannot: *accumulate* light.

But the process is neither accurate nor consistent. Each emulsion has its own bias and preferentially depicts red, blue, or green light differently from the human eye as well as from other films. When we get family snapshots developed by an incompetent photo lab, we know the colors are wrong because we recall what the correct tints of flesh, mountains, and sky ought to be. But in the case of distant celestial objects that have never furnished color to the human eye, the "true" color is anybody's guess.

Even the few deep-space objects bright enough to display some grudging tint don't allow us to determine accurately their true colors. The sky's brightest gas cloud, dimly visible to the unaided eye these nights in the middle of Orion's sword (the three faint stars dangling below the leftmost star of the belt), manages to present a weak greenish hue when observed directly through a large telescope. Yet in photographs this Orion Nebula flaunts purples and reds but no trace of green at all. We then must wonder which is more trustworthy, the visually seen green or the total lack of that tint in the photo.

The visual appearance is suspect because the eye is preferentially sensitive to the green part of the spectrum at low light levels. The lack of green in the photo is just as suspect, because that color occurs in the nebula's brighter central region, which the film may have washed out through overexposure.

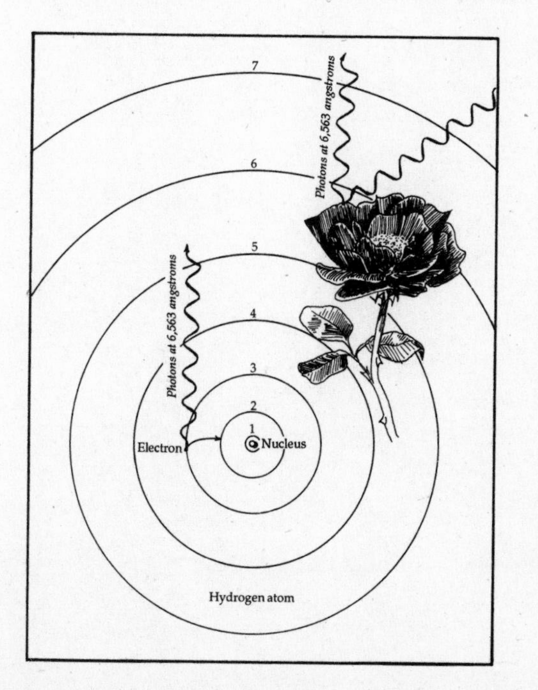

A hydrogen atom magnified 265 million times. Circles 1 through 7 represent the orbits its electron may assume. Orbit 1 is ground state, where the least energized electron will rest. When the electron absorbs energy, it jumps outward to a larger orbit. When it jumps back to a smaller orbit, it emits a photon. If the leap is from orbit 3 to orbit 2, the photon has a wavelength of 6,563 angstroms, as do photons reflected from a deep-red rose.

Neither is real. If a future space traveler trekked 1,500 light-years to the Orion Nebula, it would look different from either version, and different from any photograph ever taken of it (see color plates 1 and 2).

However, most nebulas that emit their own light are red. And not just any red. It's always the same one, a medium-deep shade that occurs at a wavelength of exactly 6,563 angstroms. This is the universe's most frequent color, for it's produced when the electron of the most common element, hydrogen, becomes excited enough to jump to a higher orbit and then tumbles down again, emitting a photon (a "particle" of light).

An electron's favorite jump is from the second to the lowest energy state, which nobody gets to see since it produces invisible ultraviolet light. But almost as frequent is a leap from the third to the second level, creating those red photons that permeate the universe. Like some infinitely vast red-light district, this particular red graces untold astrophotos, for it is found throughout the cosmos. In a universe dyed in illusion, it is an unvarying scarlet shrine to stability.

Not orange, mind you, which is extremely rare outside of some stars, and not any other shade of red either. Just that one: the incomparable crimson of our own earth's roses, the favorite color of the universe.

Realm of the Galaxies

The cab driver turns around and asks, "Where ya goin'?" It's a good question. Where are we now, and where are we heading? Around the sun? Expanding into space? Pulled by a Great Attractor? Hurtling toward Vega? Heading for a collision with Andromeda? Pick up astronomy periodicals and it's hard to make sense of it all. We might as well tell the driver "Just take us for a ride" while we look out the window, search for a local landmark, and try to unravel our destiny.

Do we live near a big intersection or out in the boondocks? If some alien spacecraft visited our section of the universe, would there be anything special about us, something that would attract notice?

Probably not—unless the aliens' curiosity had been aroused by the signs of borderline intelligence transmitted by our television signals. For our sun gleams in an outlying spiral arm of our galaxy; if *you* were to fly in the direction of some glittering city you had never before visited, would you immediately head for a nondescript suburb?

If you were a normal fun-loving alien, you'd head for the lights of town. You'd descend upon the biggest, brightest, most crowded and active region, and begin your search for life there. So, where exactly would we find the Broadway and Forty-second Street of our corner of the cosmos? If we confine our attention to our own galaxy, then the brightest region is the galactic center, floating dreamily in the southern sky on summer nights. But if our outlook is less parochial, as if our visiting aliens had come from a far-off region of reality, then the greatest gleam of light in this entire sector of the universe is . . .

. . . the Virgo Cluster of galaxies. This stupendous kingdom, almost a universe of its own, conveniently sits halfway up the southern sky these spring nights. But before you try listening for the hum of neon or searching for bright lights, be advised that this region announces itself only as a disappointingly dark district of the heavens. Its dimness, however, makes sense: In order for us to gaze into our distant intergalactic surroundings, we have to peer beyond the confines of our Milky Way, which means deliberately turning away from the rich star fields and creamy glow of the galactic plane that spills across the summer firmament. The spring sky points away from that local activity, through a sparsely occupied, thin veneer of our galaxy—an open window into the yawning emptiness of extragalactic space.

Ours is the first generation to grasp the architecture of the universe. During the first quarter of the twentieth century, many astronomers thought a single galaxy was the whole show. When the true picture came into focus we found ourselves like Rip Van Winkle, but ours was a sleep of centuries.

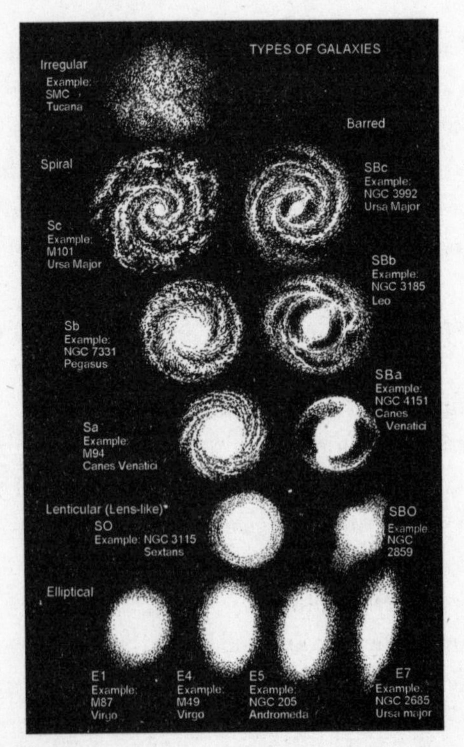

For those who like to classify things, here's how galaxies are categorized. Each is composed of many billions of suns. Our own Milky Way is an SBc.

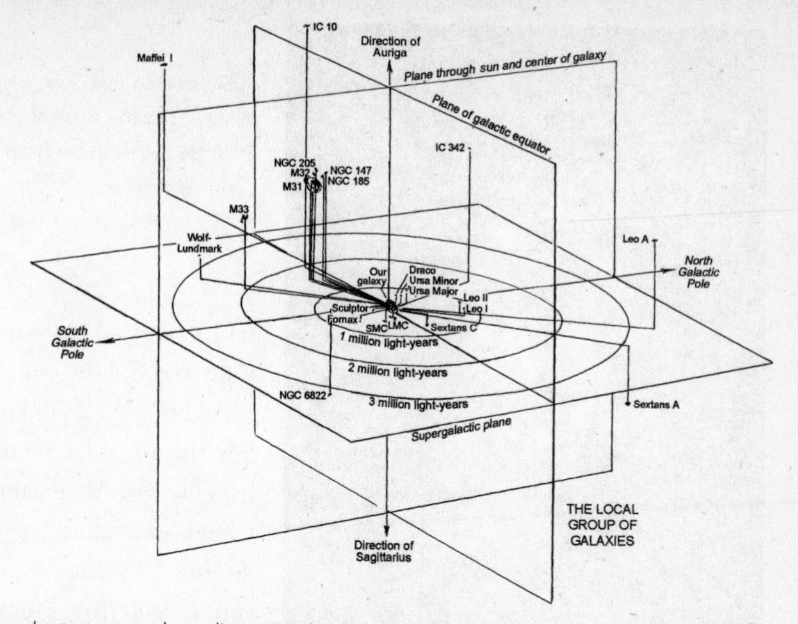

In order to portray three-dimensional space on a flat page, astronomers have set up the three planes shown here. M31 is the Andromeda Galaxy.

Dazedly taking in our surroundings for the first time, we perceived that galaxies—cities of suns—encircle us, and assume three basic forms. Spherical or elliptical ones (which generally have tremendous numbers of stars); beautiful spirals like our own, which look like frozen pinwheels; and odd, irregular galaxies of no particular shape at all, most of them midget-sized.

The little galaxies are as common as mayflies. Often, a major spiral like the Milky Way or Andromeda sails through space accompanied by several of the tiny irregulars. Most of these small galaxies are so sparse and dim that new ones have been discovered in our own backyard in just the past few decades. If it's so hard to detect those living next door, then the great open universe must be swarming with countless unseen dwarfs, and our catalogs swollen with a lopsided sample of the bigger galaxies that shine like lighthouses across the oceans of emptiness.

Our own galaxy is a member of an assembly that we call, with an epic lack of imagination, the Local Group. This small congregation of galaxies is dominated by the immense and majestic Andromeda. The second-largest member is our own Milky Way, followed by a lovely little blue spi-

M33, the Pinwheel Galaxy in triangulum is the smallest of the three spirals in our Local Group of galaxies.

ral called M33—the Pinwheel Galaxy.

Add to this trio of spirals the irregular members like the Magellanic Clouds and another two dozen dwarfs, and you've got the home team. The Local Group does more than share the same neighborhood; we're bunched closely enough to be caught in the web of our common gravity. Perhaps we even formed from the same huge protogalactic nebula (gas cloud) soon after the universe was born. Our Local Group of galaxies, some thirty or so in number, occupies a piece of celestial real estate some 4 million light-years across.

Other odd galaxies lie nearby, like a spiral named Maffei 1, discovered just a few years ago hiding behind the far side of our own galaxy and beyond the stars of Cassiopeia. (Imagine: An unknown empire of suns has always floated next door! Obscured by nearby clouds of dust, it's been hidden through all time until our instrumentation finally caught up with our curiosity.) But while such ragtag galaxies do catch our attention, nothing drops our jaws until we peer 50 million light-years away, beyond the stars of Virgo, to a cluster of galaxies with an astonishing membership of at least twenty-five hundred.

This instantly raises eyebrows: How can one cluster contain only thirty galaxies and another embrace thousands? The answer may be that we're not an independent group after all. Perhaps our Local Group is merely an outlying contingent of the Virgo Cluster!

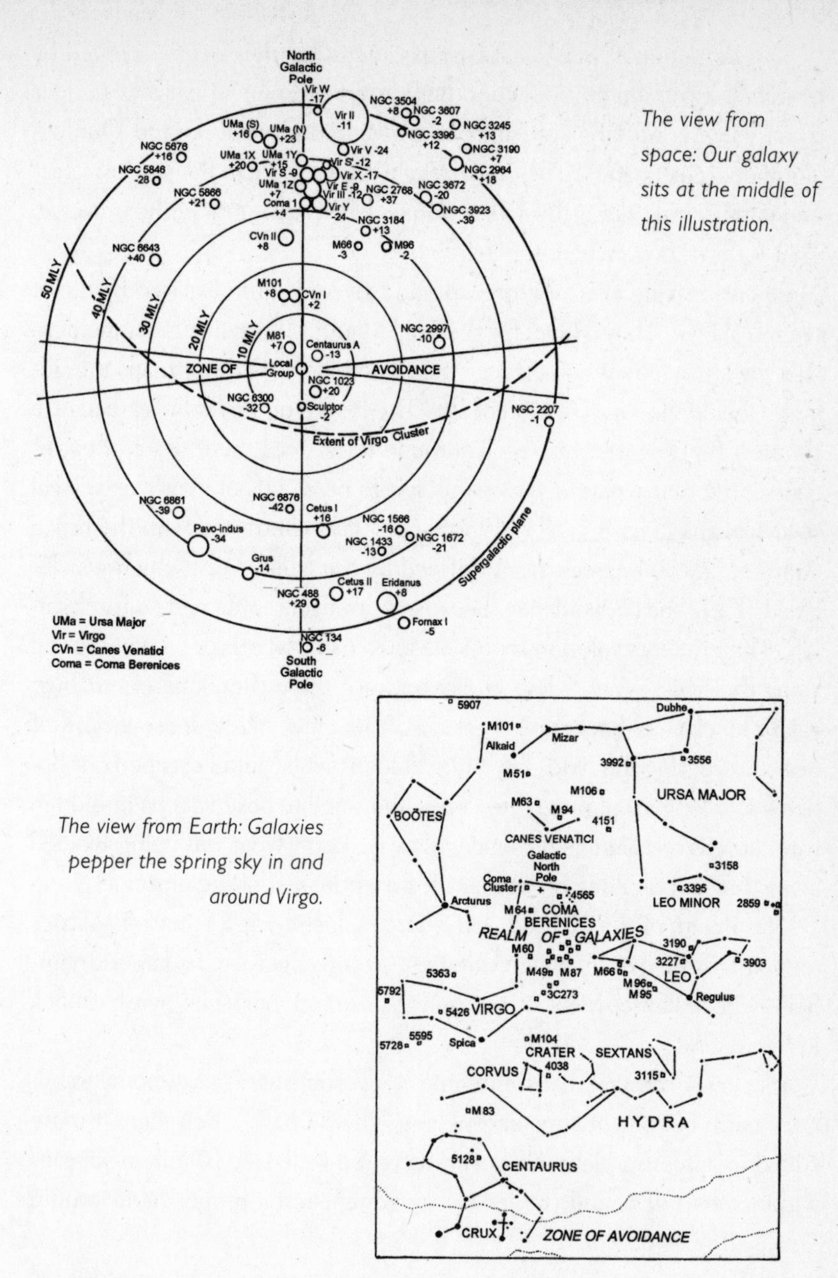

The view from space: Our galaxy sits at the middle of this illustration.

The view from Earth: Galaxies pepper the spring sky in and around Virgo.

The intriguing "Zone of Avoidance" is where the plane of the Milky Way blocks our view of distant galaxies. The least obscuration lies around the galactic poles, at right angles to the Milky Way.

This demotion of our Local Group's status has now been accepted by nearly all astronomers. Indeed, other suburban knots of galaxies pepper the region around Virgo, proving our own situation is not unique. Our loyalty, then, must shift. We belong to Virgo! *That* region, in the shadowy section of the spring sky, is the Times Square of this quadrant of the universe. That is downtown, where the action is.

So our arriving alien visitors wouldn't give our little isolated hamlet a second glance. They'd head for the bright lights of Virgo. Where, incidentally, we always knew something strange was afoot. Old star maps intriguingly labeled the Virgo section of sky "The Realm of the Nebulas" because so many fuzzy swirls were seen through telescopes there. Back then, of course, the blurry patches were thought to be clouds of nearby gas, local solar systems in early stages of formation. This contrasted with the broad swath of sky 90 degrees away, defined by the Milky Way, which was labeled "The Zone of Avoidance" because it strangely contained *no* fuzzy spirals. The mystery's solution is now obvious: External galaxies are not found along the Milky Way region of sky because there the plane of our own galaxy blocks any background galaxies from view, like a dense swarm of bees obstructing the fields beyond. Which is why Maffei escaped our notice for so long, until radio telescopes and satellite observations at hidden wavelengths revealed its presence. Now we know what was being avoided along that cryptic Zone of Avoidance: the entire rest of the universe!

The Realm of the Galaxies—the Virgo Cluster—is an atlas of galaxies randomly strewn every which way, face on and edgewise. Its largest member—whose name, M87, is not exactly inspirational—sits nearly smack in the middle.

It's almost comical that something as awesome as a galaxy, home to billions upon billions of suns and planets, should have a designation more suited to a license plate. However, there are well over 10 billion galaxies within view today, and the job of assigning each a proper name would quickly lose its charm.

M87 is among the largest galaxies known, a gargantuan elliptical blob that will never win a beauty award. Still, it's the largest *ball* in the known universe. It consists of several trillion suns, give or take a few, and probably the same number of planets. This means that if you were given the task of counting its stars—not visiting them, or cataloging

The colossal elliptical galaxy M87 at the center of the Virgo Cluster. The smaller spheres are some of the fifteen thousand globular clusters that surround the bright core of the galaxy like a swarm of fireflies.

their characteristics, but merely counting—and managed to run through ten every second, the assignment would last from the last ice age until the present time.

An enormous black hole sits at the center of M87. Thousands of globular clusters of stars, each with hundreds of thousands of suns, orbit around it. M87 is virtually its own complete cosmos. To any of its possible inhabitants, it would be *more than could be explored, ever.* The rest of the universe, including our Local Group, would seem utterly irrelevant, thoroughly unnecessary.

Spraying out of M87 is a fiercely violent double-jet of material rushing at nearly the speed of light. From this bizarre dual structure comes radiation and energy of such power that it can be loudly heard (with radio telescopes) across the gap of 50 million light-years that safely separates us from it. Its origin may be that colossal black hole hiding in the galaxy's core.

M87's central black hole is no mere collapsed star. Its mass has been calculated to equal somewhere between 10 million and 100 million suns. Because such a heavy black hole does not, paradoxically, produce the crushing tidal forces of smaller ones (because it is initially less compact, which dilutes the effect), it could possibly be a navigable tunnel to another time or a distant place in the universe. A pilot with the skill and

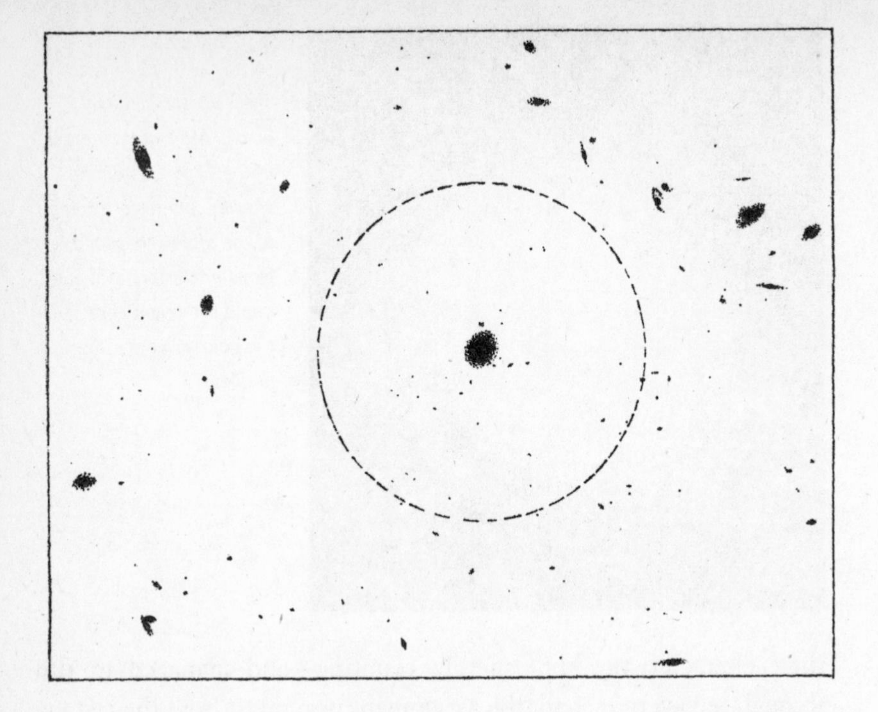

M87 sits heavily, with the mass of twenty Milky Ways, in the middle of the Virgo Cluster of galaxies. M87's halo of hot gas (at 30 million degrees) extends out to the dashed circle—1 million light-years in diameter!

courage to hurl a spacecraft across its event horizon (see page 103) might just make it through this alien underpass to regions lying on the other side of our imaginations.

But no one will ever get the opportunity to try. If we someday gain the ability to travel at virtual light speed, the time required to voyage to the Virgo Cluster would equal the span that has elapsed since the age of the dinosaurs. Even the brightest galaxies of Virgo are too distant to appear without optical aid. Binoculars barely display them as hazy blobs. Detail reluctantly springs forth only through larger amateur telescopes, and then just on a few distinctive members such as the famous Sombrero and Black-Eye galaxies. Nor has the distinction of being the nearest major galaxy cluster been enough to bring Virgo to public attention. So it sits, out of sight, out of mind, floating invisibly in the dreamy depths of night.

When we look still deeper into space, to more distant galaxy groups, their members appear too small and faint to display significant detail of any kind, through any instrument. It may sound impressive to say that a galaxy lies, say, 500 million light-years away, but the greatest scenes of galactic splendor originate in our own region of space. The stunning recent photographs of galaxies reproduced in magazines or coffee table books are almost invariably those of Virgo, the Local Group, and the galaxies scattered in between.

While members of the Local Group have taken gravity's vow to stay married through all of time, the Virgo Cluster lies far enough away for the intervening space to be stretched by the **Hubble Flow,** one of the names for the expansion of the universe. Because the cosmos is growing larger, those thousands of galaxies are bidding us adieu at the rate of 110 miles per second. If we wished to launch a rocket to visit one of the Virgo galaxies, we'd have to tack on an extra 110 mps just to keep pace with the stretching of space between us.

This is something straight out of a funhouse, or the kind of dream you get from eating too much pizza before bedtime: You try to run down a hallway, but it stretches ever longer before you as you move your legs. Since our best rockets now take us just one-tenth that speed, we have a long way to go before we can even keep pace with the expansion of the universe in our own narrow corner of the cosmos.

The "black eye" of galaxy M64 in Coma Berenices is caused by vast clouds of dust across its bright center.

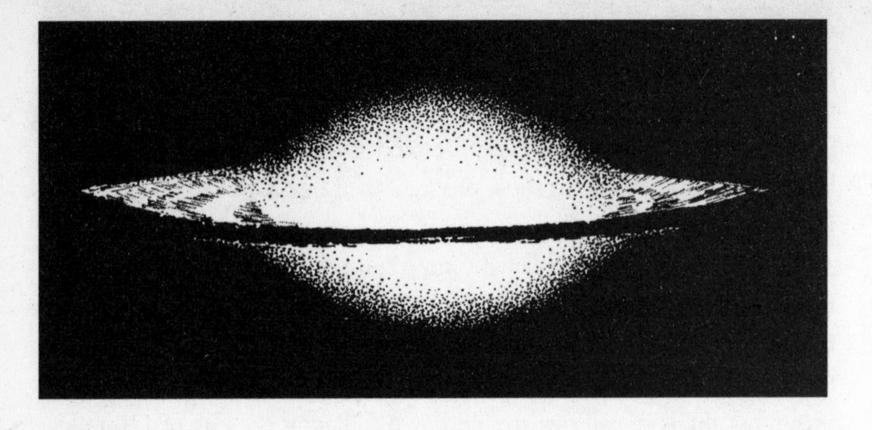

The Sombrero Galaxy is a tightly wound spiral some 60 million light-years away, in the direction of Virgo.

This carries us to the other half of the question we pondered in the taxicab: Where are we going, and how fast? Absolute motion was shown to be an invalid concept a century ago. And since there's no constant grid in space to tell us where we stand, we are always forced to express our velocity relative to something else.

For instance, we know that Earth circles the sun at 18 miles per second. But when people then assume they're flying through space at that speed and will return to their starting point after a year of orbiting, it's simplifying things to the point where fiction replaces fact. For openers, our planet is also moving, together with the sun, toward the bright star Vega at about 12 miles per second (44,000 miles an hour), so that we never really "complete" an orbit in the sense of returning to where we started. Our path through space is not a loop but a spiral, like an endless bedspring!

Additionally, we're on an even faster journey around the center of the Galaxy at almost 200 miles per second. This is a frenzied, major motion, some four hundred times faster than a rifle bullet. Yet it's easiest to overlook because all the stars around us fly along together in the same direction. This rapid galactic ride experienced by our entire celestial neighborhood does not, therefore, greatly alter the appearance of the night's constellations over the centuries. We generally ignore it and instead focus on small variations within that stream. The trivial 12 mps amble toward

Vega is just a little side current in the big forward flow—and yet many texts merely announce we're heading toward Vega, period, as if that were the end of the story.

Just within our galaxy, then, our lively planet is performing three major motions. And then, additionally, there's the odyssey we're on as a part of the Galaxy itself as it rushes toward Andromeda at some 50 miles per second.

Or are we stationary, with Andromeda coming toward us?

Here we reach the point where we cannot declare who is it that is moving. We can't speak definitively of our galaxy's motion since there is no adequate outside reference point. A third object is always necessary to describe the motions of two others, just as two are needed for the movement of any *one* to make sense.

Not clear? Consider: If Earth were alone in the universe, it would be meaningless to speak of its motion. Attach rocket boosters to its underside and speed it up, and Earth still travels just as fast—namely, not at all! For how can motion through nothingness carry any meaning? But add a second object (let's say the moon) and suddenly you could say Earth is zooming toward the moon at such-and-such a speed. Notice, however, this is not enough reference to declare who exactly is moving. Earth could be stationary and the moon moving, and you couldn't tell the difference. It can only be said, "Earth is moving at *x* velocity *relative* to the moon." To determine who was in motion, Earth or moon, you'd need at least one further body to establish an outside reference point.

That's why we simply say the space between us and Andromeda decreases by 50 miles each second. We cannot say who is moving toward whom unless we use the other galaxies in our group as a reference. But there are too few of them, and they're all in motion as well, so the whole thing gets shaky.

The closest we can come to perceiving "absolute motion" is by viewing irregularities in the background radiation of the Big Bang, which permeates all space. Its 2.735-degree temperature (that's in the Kelvin scale, marking the degrees Celsius above absolute zero, where all motion freezes) shows a slight upward and downward shift in opposite regions of the sky. This must be due to our own deviant motion relative to the overall expansion of the universe.

At last—an "outside" marker! It indicates we are flying toward the constellation Hydra, very near Virgo itself. We—and the entire Virgo Cluster—are thus being yanked askew from the Hubble Flow by an unknown supermassive *something* lying on the far side of the Virgo Cluster. The reality of "the Great Attractor," as this something is known, has been called into question, and as of the mid-1990's it remains either a giant object or a giant mistake.

But even here, we can get a sense of our deviation from the smooth expansion of the universe, but not learn where we are in any absolute sense. Nor should the unreality of absolute motion seem mysterious. When on a plane, for example, you avoid running into the harried flight attendant with the cart because you focus on just your motion and hers, without bothering to calculate each of your trajectories through space. If you had to factor in your *true* motion (the 500-mph speed of the plane, the 66,000 mph of Earth, and all the rest), the movements and events leading up to your knocking the diet soda all over the passenger in 16C might seem impossible to calculate, let alone prevent. Local deeds relative to some object of momentary interest are all that matter on Earth as in the heavens.

To top it off, on the deepest levels of reality, nobody is going anywhere. Relativity tells us that distance mutates depending on one's velocity, so that even the most remote galaxy becomes just a city block away if one is traveling just under the speed of light. We may gaze at the stars and be awed by their distance, but the distance between objects will readily shrink like Alice in one of her adventures, given changing circumstances like gravity and speed.

If you measure the space between yourself and the far side of the room, the yardstick might read 14 feet. But if you were traveling at the perfectly permissible speed of 184,500 miles per second (99 percent of light speed) and performed the same measurement the instant you reached the same spot you now occupy, the tape would read 2 *feet*! No inexplicable process distorted your measuring device; the room is now actually 2 feet long. Length and distance are changeable.

At the speed of light, you'd find yourself everywhere in the universe at once! In an ultimate sense—if one had the same perception a photon of light would have—there is only one place in the cosmos, and that place is simply "here."

So, yes, there *are* local landmarks, the biggest of which is the army of galaxies arrayed behind the stars of Virgo. Our *place* in the cosmos is fairly clear. But as to our motion and our destination, we can only say we are going nowhere—fast!

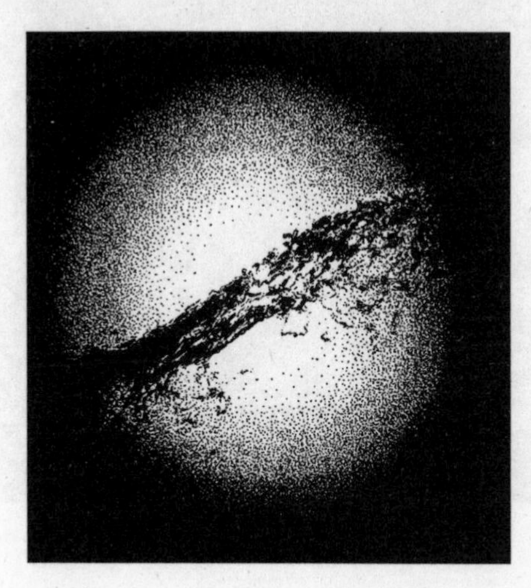

It's rare that a dust lane girdles the nucleus of an elliptical galaxy, but that's only one mystery about Centaurus A. It also has two giant lobes of ionized gas extending across 2.5 million light-years. They could engulf both Andromeda and our galaxy if Centaurus A were between us.

Mysteries of Jove

Even if we thoroughly enjoy immersing ourselves in such conceptual quagmires as curved space and fifth dimensions, there is a special magic in the actual night sky. A quick step outdoors finds the soothing stillness of the starry dome, and suddenly there's something simple: the brightest "star" of the midnight heavens—the planet Jupiter. But there's more here—much more—than meets the eye. For, with every robot emissary we send its way, Jupiter answers by teasing us with ever deeper mystery. Once thought to be a simple case of a colossal ball of hydrogen and ammonia, that gargantuan living orb of swirling color increasingly awes and overwhelms us.

Look at it. Nothing is as brilliant except for the moon and twilight-hugging Venus. Guarded by powerful magnetic fields that surround it with lethal radiation, Jupiter is conspicuously easy to find, but mysterious as a dolphin's dreams. No wonder the ancients deemed it "king of the gods," a status bestowed by the Greeks as well as the Romans, for whom it was the almighty Jove. Other than believing it had absolute power over their lives, they didn't even try to assign it attributes. Jupiter was the ruler of everything, period. And these ancient civilizations weren't far off the mark: Modern knowledge has confirmed Jove's royal rank.

Jupiter is so large that thirteen hundred planet Earths dropped inside wouldn't quite fill it. It has more substance (mass) than all other planets combined—and doubled. Our solar system is essentially made of the sun and Jupiter. All the rest is an afterthought.

But size alone, even in our superlative-loving culture, wouldn't qualify it as one of the genuinely amazing treasures of the Museum of the Night. It also brims with power, the kind of raw gravitational influence that wrenches moons and comets, flinging them off in careless new directions. And internal power, too—a maelstrom of magnetic fields and continuous lightning buzzing with lethal intensity and unknown effect. And beauty enough for a dozen Uranuses or Neptunes. Few astronomers would hesitate to call it the most awesome planet in the known universe.

To explore it with our minds, we should first find it in the heavens, a task as simple as spotting the brightest "star" in the midnight sky.

Jupiter at Its Best

The dates when Jupiter will be brightest, and highest at midnight. It will be prominent from one month before to four months after these dates.

1995: June 1	2002: Jan. 1
1996: July 4	2003: Feb. 2
1997: Aug. 9	2004: Mar. 4
1998: Sept. 16	2005: Apr. 3
1999: Oct. 23	2006: May 4
2000: Nov. 28	2007: June 5

When we look at the table above of Jupiter's brightest dates, the simple pattern is obvious at once: Jupiter reaches its greatest prominence one month later each year.

You can turn up the juice several notches by pointing ordinary binoculars its way. Lying a mere 400 million miles from us, Jupiter is close enough to display a disk, a shape, even when viewed through the cheapest field glasses. Such a binocular-induced transformation from a dot to a distinct disk unfolds with only two of the universe's objects; the other is Venus. All of the thousands of stars and few remaining planets remain mere points of lights. (Well, Saturn looks *elongated* through binoculars, but the rings are not recognizable.)

Jupiter also displays a startlingly obvious retinue of moons lined up like track lights. Each night the four little Jupiter-hugging dots change their orientation: three on one side of the planet and one on the other, or two and two, and other possibilities in an unceasing minuet that is so simple to observe that it's easy to forget their importance in human history.

Galileo, using a truly pathetic low-power telescope, peered at Jupiter in 1610 and instantly revised the universe. The satellites conspicuously circling the brilliant planet offered the very first proof that Earth is not the center of all motion. In an instant, thousands of years of philosophical debate came to an end.

Or at least it should have. While Galileo naïvely assumed that such vivid evidence would carry the day, he soon learned that most people, then as now, are not philosophically agile, not quick to jettison long-held opinion. In one of the most frustrating moments ever recorded, priests peering into his eyepiece a few years later would comment only, "I see nothing!" Faced with revising the Earth-centered doctrine that had bewilderingly been made a religious principle, they pretended they could not see a thing.

Sizes of moons and planets overlap. Jupiter's Ganymede is the largest moon in the known universe.

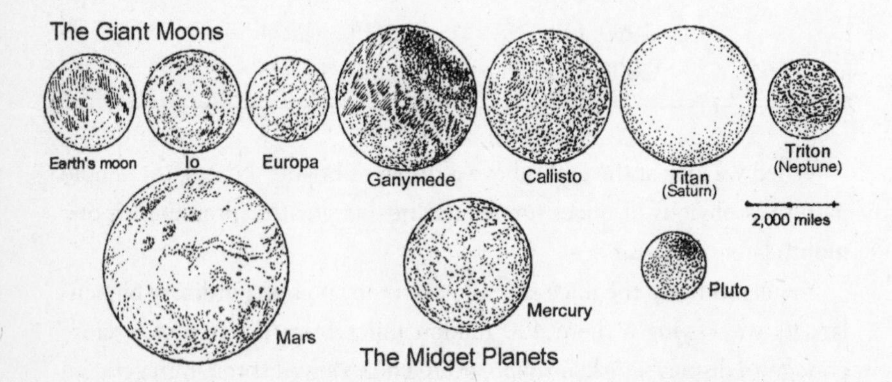

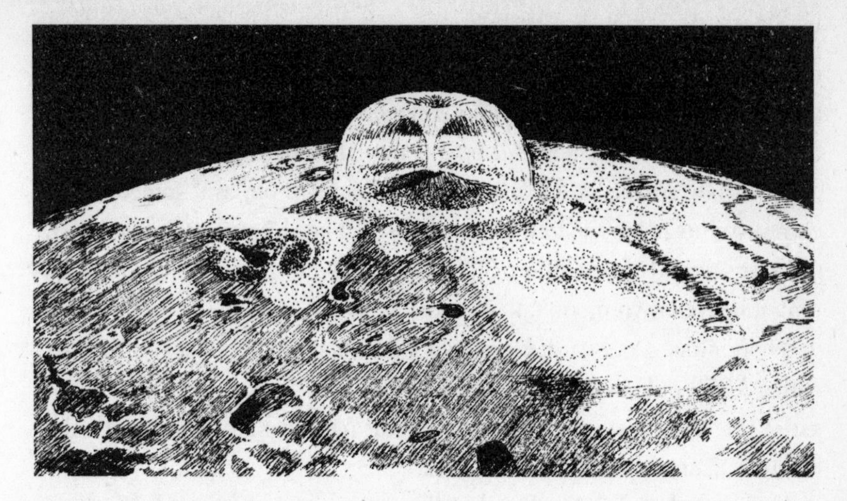

Jupiter's magnetic and tidal forces cook the insides of Io until they erupt through the crust, strewing the surface with various forms of sulfur. Shown is the volcano Prometheus on Io's equator.

Today you can repeat Galileo's mind-numbing discovery without even owning a telescope. Simple binoculars provide far greater clarity than Galileo's first instrument, whose images were wildly smudged with false color. Opticians had not yet discovered the simple trick of combining two types of glass to reduce the propensity of lenses to focus each of light's component colors in a different spot, smearing the image. Today's optics offer a vestigial taste of this chromatic aberration when they produce yellow or purple fringes around bright objects. But in the seventeenth century a peek through any telescope was a psychedelic experience.

You can do more than merely *see* those four little "stars" hugging Jupiter. (There are only three? Then one is temporarily in front of, or behind, Jupiter. Stick around and it'll emerge.) You can really impress your friends by announcing them by name.

Identifying the individual moons of Jupiter might seem an impossibly difficult, esoteric enterprise. But getting the hang of it takes only a minute or two.

The inner one, Io (say it either way, EYE-oh or EE-oh), zips around Jupiter in just 1.77 days, changing position while-U-wait. If you're viewing Jupiter through even the smallest telescope, the movement of Io over the course of an hour is obvious. This is the satellite on which the fabulous

Voyager spacecraft found ten erupting volcanos—meaning Io has the most mutating, active surface in the known universe. Its slightly orange color comes from stygian sulfur in various stages of solidification.

The next nearest moon, Europa, is the smoothest known body in the cosmos. The streaks and markings appear painted on, as if someone had commissioned a moon-sized mural of abstract art. Europa orbits in exactly twice Io's period of revolution, 3.55 days—still impressively fast when you remember our own moon takes almost a month.

Then, moving outward: Ganymede, the solar system's largest satellite, with a 3,279-mile diameter. You can quickly spot it because it's the brightest of the four. Watch it make one circuit per week, a period twice the length as Europa's and four times Io's.

How wonderfully peculiar that these three march around their parent world in such a simple 1:2:4 orbital relationship, each taking twice as long as the next satellite inward. It means that the moons are resonating with each other, gravitationally locked in sync like a military drill team.

Finally, farthest from Jupiter there's giant Callisto. Like Ganymede it's mostly ice, plain water ice. Dig down a thousand miles and you're still mining ice, perhaps hinting at a potentially lucrative Sno-Kone industry in the twenty-second century. Callisto is just a whisker smaller than Ganymede and both are larger than the planet Mercury, or Pluto. All four of these giant Jovian bodies are collectively called **Galilean** satellites in commemoration of their discovery four centuries ago.

Remember, the apparent order of these four moons can change as they revolve, so while Io usually appears innermost and Callisto farthest out, Callisto can sometimes seem the nearest to Jupiter when it's about to pass in front of, or behind, the planet. That needn't throw us for a loop: Io is always fastest, Ganymede brightest; and if one seems quite far from Jupiter, then it's Callisto rather than Europa. A few moments' detective work does the job, and your family suddenly looks up to you as a Newton who can identify the moons of another world.

A dozen other Jovian satellites have been found, but they're faint chunks of debris invisible to amateur telescopes—not in the same league as the four Galilean heavyweights parading in front of mere binoculars half a billion miles away. The five smallest, most recently discovered satellites, each an irregular chunk of debris that would fit within the border of Rhode Island, were detected only by *Voyager*'s close-up snooping.

Were it not for their eternal closeness to Jupiter's dazzling glare, the Galilean globes would all stand out clearly without any optical aid at all! In other words—and how's this for trivia—not one but five moons are bright enough to be seen in our sky with the naked eye: our own, plus the Jovian quartet. Sharp-eyed persons, especially children, have seen them for centuries. A typical report is of a father gazing through the family's small telescope while telling his seven-year-old that when it's her turn she should look for the little dots "to the right of Jupiter." To which the child, staring into the sky, says, "No daddy, they're on the left!" (Telescopes flip the image.)

Back when Britain ruled the seas, the Galilean satellites served as an ingenious clock and navigational device used around the world. It was one of the primary functions of England's Royal Astronomer to calculate when they'd reach prominent configurations; when, for example, all four would bunch up on one side, or two would come so close together as to appear to merge. Then British mariners on the high seas, using their hand-held spyglasses and invaluable (and secret!) tables obtained before setting sail, could, by observing Jupiter, figure out what time it was, since accurate knowledge of the time of day was required to calculate their longitude. For over a century people found their way around our planet by watching the moons of another.

British naval officers check the position of Jupiter's moons against tables provided by the Admiralty to determine Greenwich mean time before the invention of the chronometer.

It had always been noticed that Jupiter's moons sped up for half

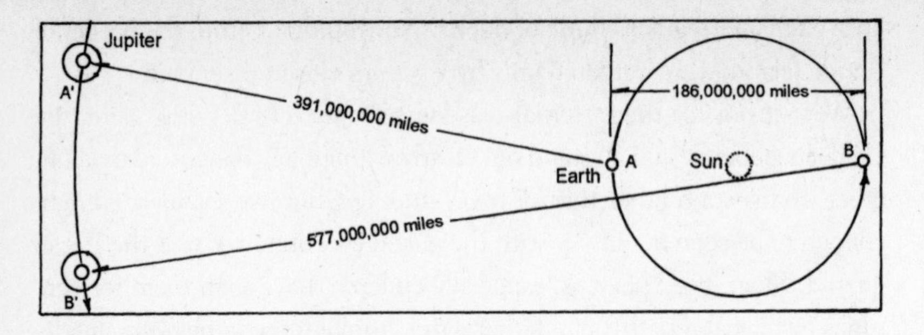

In the six-month period during which Earth travels from A to B and Jupiter from A′ to B′, the distance between the two planets increases by 186 million miles. The light from Jupiter's moons takes sixteen minutes longer to reach us at B, causing their motions to appear delayed.

the year and slowed during the other half, a bewildering trait that was duly incorporated into the navigation tables. The satellites would arrive at their predicted positions ahead of schedule whenever we approached Jupiter, and trail behind when Earth was moving away. What was going on?

During the seventeenth century Ole Roemer, a Danish astronomer, correctly deduced that the advance or lag had to do with the speed of light. The images from the Galilean satellites had a shorter distance to travel to our eyes when Jupiter was nearer, and farther when Jupiter was on the distant side of the sun. The latter situation caused their light to arrive sixteen minutes "late" and produced the foot dragging in the moons' ballroom steps. Using this reasoning, he correctly calculated the speed of light to an accuracy of 25 percent, by far the earliest reliable estimate of light speed. It took more than a century after his death to improve on it.

If the mere moons of Jupiter changed our view of the universe, aided us in earthly navigation, and revealed light's secret speed, what then do we find when we turn our gaze to the parent planet itself?

The naked eye shows only a dazzling, creamy white "star" that displays little willingness to scintillate. The old saw about being able to tell stars from planets because the latter never twinkle is advice with merit. Stars appear one-dimensional—mere points of light whose fickle beams are easily deflected by passage through our atmosphere. Planets, though seeming starlike, have size and dimension; they're a broad stream of numerous light beams—less susceptible to atmospheric legerdemain.

Jupiter's equatorial bulge is perceptible on this scale—the equatorial diameter is 6.8 percent larger than the polar diameter, and it's 11¼ times Earth's (lower right).

The vision of Jove improves with ordinary binoculars, and becomes a thing of wonder though even the cheapest telescope. Any instrument reveals Jupiter as a squashed disk, its equatorial diameter noticeably wider than its polar dimension. This oval shape comes from a wild, runaway-train kind of rotation: The giant planet spins in just nine hours, fifty minutes. Powerful stuff, that the largest planet also boasts the solar system's fastest spin. Its equator races along at a frenzied 28,000 miles per hour (twenty-five times faster than ours!) making for an amusement park ride of centrifugal force that more than offsets Jupiter's stronger gravity. This wild equatorial spin stretches out Jupiter's fluffy, barely-denser-than-water composition by an extra 6,000 miles, an exhibition of power that arrives with the first telescopic peek.

The smallest telescope also shows belts or bands running parallel to the equator, like the markings on the belly of a bumblebee. Better instruments and steady observing conditions (when stars are not twinkling) bring a world of intriguing detail. Dark and white ovals, swirls, festoons—an arabesque of endless complexity makes Jupiter the most detailed world visible to Earth's instruments. There's also a single grayish-pink, oval patch. . . .

The Red Spot! Has there ever been a less inspired name—for anything whatsoever? This ancient feature, sometimes conspicuous, sometimes impossible to find, looks like an elliptical pink ocean. But Jupiter doesn't have a surface. So the phenomenon must be a kind of storm, a cyclonic

Clouds on Jupiter show patterns of cyclones and eddies familiar to weather watchers on Earth, but the scale is different. Two planet Earths would fit neatly into the Red Spot, at right.

revolving whirlpool like an eddy among the rapids, caused by the turbulence of two nearby streams of material rotating around Jupiter at different rates.

Seen for centuries, the Red Spot is a tempestuous province floating in the atmosphere like a colossal hurricane twice the size of Earth. Seething with lightning a billion times more intense than our own planet's version, the stygian environment of the Red Spot (and indeed the entire massive planet with its other swirling, ever-changing formations) seems like a bewildering, complex painting in continual motion. All of its markings, spots, ovals, and curlicues, built of such hydrogen compounds as methane and ammonia, rush along like frenetic New York taxis, for Jove's snappy rotation makes it a place where everything's in a hurry.

When we shoot electrical sparks through a jar containing the materials in Jupiter's atmosphere, all the amino acids—life's building blocks—are created within a week. What, then, may lie beneath those colorful, frozen cloud tops? Could 4 billion years of potent chemistry have cooked up creatures swimming serenely through the soupy liquids of the lower levels? The *Galileo* spacecraft, scheduled to orbit Jupiter and even penetrate the Red Spot in 1995 and 1996, is not adequately equipped to provide answers about Jovian life. The question will remain a mystery at least until the early 2000's when a more sophisticated robot may brave the fierce pressures and intense lightning.

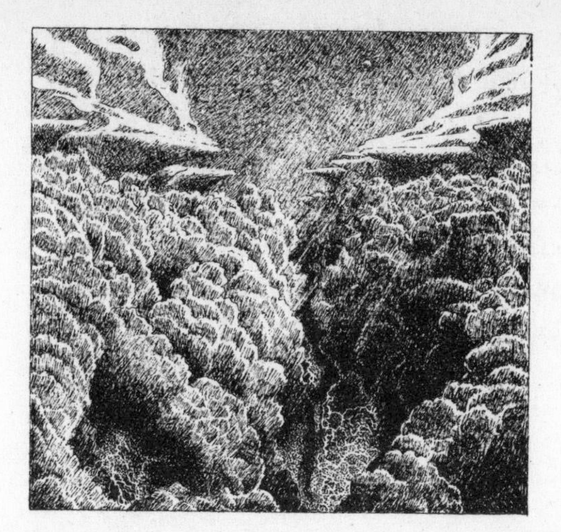

Jupiter has lots of atmosphere! The upper strata of clouds are ammonia crystals. The lower deck is ammonium hydrogen sulfide tinted by molecules of sulfur. It is often stormy—with thunder and lightning and high winds.

Jupiter, of course, grabbed headlines in July 1994 when it was mercilessly pounded by twenty-one fragments of the disintegrated comet Shoemaker-Levy 9. By some curious cosmic choreography, the powerful impacts coincided with the *Apollo* moon landing's silver anniversary week like celebratory fireworks, creating temporary dark features detectable with amateur equipment. This was the planetary event of the century for backyard astronomers. The comet was even thoughtful enough to choose the time of year when Jupiter would be visible at convenient viewing hours around the world.

While the Jovian features we see from Earth are merely the frozen upper layers of the clouds, temperatures underneath them rise continuously. Unlike Earth, Jupiter doesn't need an outside source of warmth. It generates twice as much heat as it gets from the sun, an enviable independence bequeathed by its legacy as a stellar also-ran. Had it one hundred times more mass, then pressures and temperatures in its core might have ignited the nuclear reactions that define starhood. As it is, Jupiter's core got hot, but not hot enough to begin nuclear fusion. Still, it's impressive by planetary standards: Jupiter's heart is the most fiery place in the solar system outside of the sun. The warmth rises upward, guaranteeing the existence of a zone in Jupiter's thick soupy environment that is a comfortable room temperature.

Although there's no evidence to support it, one can easily imagine life forms swimming or floating there, fed by the carbon we know is present, and not bothered at all by the scarcity of oxygen. (Even on Earth there are anaerobic bacteria that do not require oxygen. All life as we know it, however, needs carbon.)

To the naked eye, binoculars, or telescope, Jupiter remains the award winner for planetary observers. It's worth buying a telescope for this giant world alone. Try it out. If you're not satisfied with the view, return the instrument. Galileo had to deny what he saw for fear of death. The refund policy at your store is probably less intimidating.

Everything You Never Thought to Ask About the Crescent Moon

Only a handful of celestial sights have no analogue in earthly experience. The rings of Saturn, for example, were so unexpected that Galileo never quite figured out what he was seeing. To his dying day his best drawings depicted them as "handles" attached to Saturn, as if he were sketching a sugarbowl. There simply was no example on Earth of a globe surrounded by unattached rings.

The same is true of the crescent moon. We can easily create a synthetic crescent by viewing a ball illuminated from behind. And while there are common crescentlike shapes (the French croissant, bananas, and some beaches), a true crescent brings to mind the moon alone.

Even through the most powerful telescopes, only two other objects in the entire universe—Mercury and Venus—can assume a crescent shape. This scarcity is solely an artifact of our location, as if we asked the Heavenly Realtor for a quiet, minimally crescented neighborhood. But other celestial precincts present a radically different night sky. If we lived across the tracks on Jupiter or Saturn, we'd be able to stare up at a fantastic array of more than a dozen various-sized crescents, the fabric of fantasies. The single naked-eye example seen by earthlings is a peculiar situation. For it must be a common condition throughout the universe to see spheres (nature's favorite shape) illuminated from the side turned away from the observer. There are surely worlds that have never beheld a cube or pyramid—but the crescent is a universal motif out to the farthest galaxies.

One would think the crescent's more or less solo appearance in our world would grant it special status. Well, yes and no. On the one hand the crescent moon does appear in many early cave paintings and, together with a brilliant "star" that is probably Venus, emblazons the banner of Islam. But as far as college courses on astronomy are concerned, it is rarely discussed or thought about. Small wonder its comings and goings are as cryptic as an infant's smile of the same design.

Very few of us, for example, are aware of the simple fact that the moon appears as a crescent whenever it lies closer to the sun than we do. This goes on for almost half the month.

About a third of those crescents, however, are impossibly difficult to see. The slimmest ones stand almost in line with the sun and therefore set around sunset. Such ultrathin phantoms lurk in secret near the skyline, buried in bright twilight. Like other sun-hugging apparitions, they've set by the time the sun drops much below the horizon. So while cartoonists often portray midnight skies with a slender crescent moon, such a sight is impossible. A thin crescent cannot appear in the middle of the night.

For the crescent-crazy we offer Saturn. Here nine moons present disks large enough to be distinguished as crescents. Not enough? Get out your telescope and watch for the other eleven moons, or direct it at the rings: They are made up of billions of tiny crescents. (The near moon is Mimas.)

The lunar phase that notifies us that crescents will soon arrive in our lives is new moon, which officially starts each lunar cycle. New moon really means no moon; it's an invisible nonphase (unless its ebony disk passes directly in front of the sun to produce an eclipse). But outside astronomical officialdom, many around the world define "new moon" as the first appearance of the thin crescent a day or two later. Certainly it is the tradition of the religion of Islam to begin the month, and many holidays, with a visible new moon—such a sighting is a main purpose for minarets.

This first emergence of the moon after its several-day absence is that of a delicate eyelash-thin arc deeply immersed in twilight glare, a challenging target. Perhaps that's why millions of people, most of them not Muslims at all, enjoy the esoteric hobby of trying to spot this "youngest" possible moon.

You may never have heard of this peculiar if harmless hobby. But astronomy magazines routinely publish letters and articles dealing with claims of finding ever-thinner moons. Sharp-eyed spotters in the Middle East's crisp conditions probably do even better, without reporting their observations to Western journals.

Every sport has its numerical criterion. In golf it's strokes; for runners, seconds of time; in chess, points. Crescent watchers care about the "age"

The moon moves eastward in its orbit around Earth 13 degrees per day, or about its own diameter each hour. The size of our moon has been exaggerated here to show how the crescent changes appearance. The moon's actual diameter is shown at upper right. The crescents indicate the moon's positions at nightfall for the first four days after the new moon.

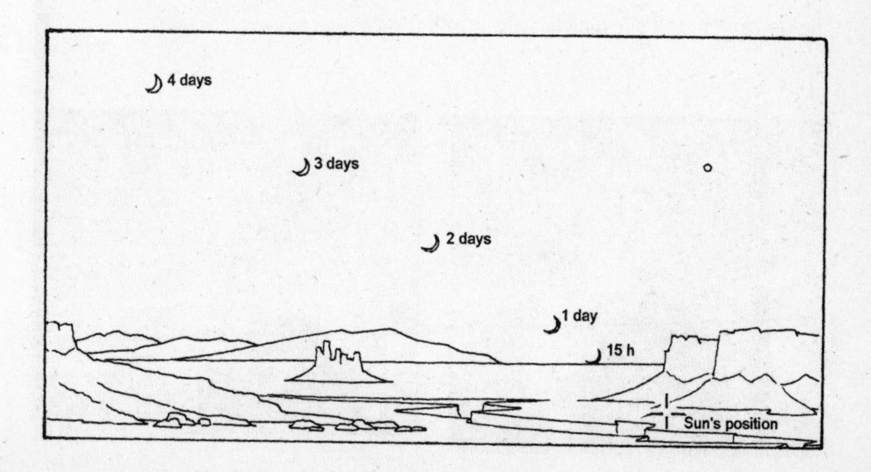

of the moon. This is simply the time elapsed since it was new, or most in line with the sun. Everyone agrees that a two-day-old moon is easy to spot. It's generally also conceded that a moon younger than about 14 hours is impossibly difficult. A 24-hour moon is challenging, but often observed. Somewhere between 15 and 20 hours, then, lies the true challenge, the monthly hunting grounds for crescent fanatics around the world.

Binoculars make the task so much easier that the real enthusiasts divide the contest into two arenas, those using them and the purists employing just the naked eye. The record with binoculars is 13½ hours; nobody has yet been able to find a moon younger than 15 hours without them. In temperate regions, the only realistic chance of spotting a very young moon in the evening sky occurs in February, March, and April, when the moon's orbital path makes a steep angle with the western horizon. Only then will the 13-degree separation between sun and day-old moon stretch straight over the sunset, not splaying to the left to immerse the hair-thin crescent in horizon haze.

This changing angle of the moon's orbit with respect to the horizon is also responsible for the different orientations of the crescent itself. The crescent lies "on its back" like a boat following those same February-through-April sunsets in most of the developed world. The rest of the year it appears on its side, like an archer's bow. From equatorial regions the moon's path forever makes a near-vertical tilt with the horizon, so that a smiling crescent is the unvarying logo of the tropics. Conversely, a sideways crescent is the only moon ever seen from polar areas.

The crescent moon's orientation changes with latitude.

The only time you'll see a frowning moon is in broad daylight.

In short, you can tell your season or location from the slant of the crescent moon.

The moon's an eternal optimist. Never, anywhere in the world or at any time of night, does it display its bright side upward as if frowning, despite often being depicted that way by artists. If its personality has a sad side, it's unmasked only in broad daylight, as if unhappy with the sun's competition. A frown can appear only against a blue sky, when the moon's seen within a few hours of noon.

Dawn, too, has its own conditions, separate from those of the evening sky. Here the situation is reversed, because the only time a morning crescent will look like a smile in our part of the world is not spring, but August through October.

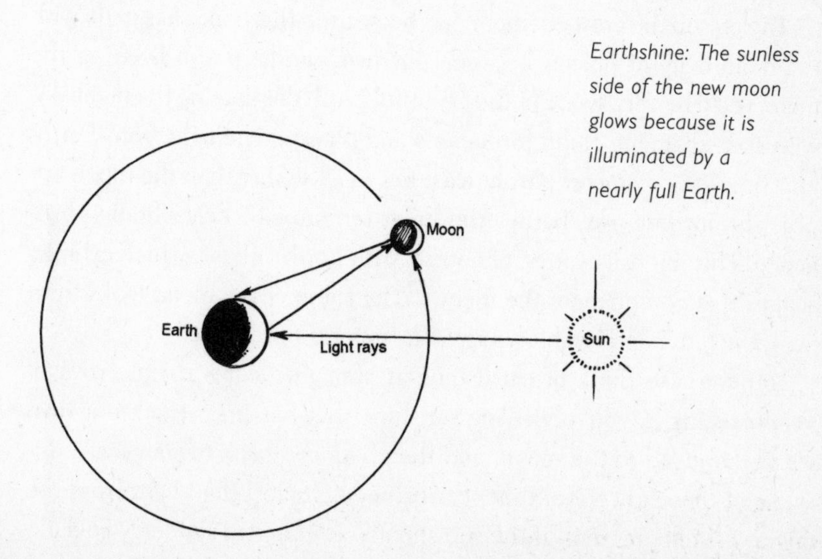

Earthshine: The sunless side of the new moon glows because it is illuminated by a nearly full Earth.

Earthshine lights the sunless side of the moon early in the first quarter and late in the last quarter of the lunar month.

As evening twilight darkens and the crescent moon brightens, its dark unlit portion glows supernaturally. This eerie phenomenon, traditionally called "the old moon in the new moon's arms," is now simply termed **earthshine.** That's just what it is—a portrait of our own planet's light reflected back to our eyes.

Of the universe's trillions of celestial bodies, the moon alone is near enough to our world to serve as a mirror, to bounce back our own brilliance for our narcissistic enjoyment. This dependable characteristic occurs only when the moon is a crescent. Then its dark portion shines so distinctly that markings on its unilluminated hemisphere stand out clearly. They're even more dramatic, if a bit spooky, through binoculars.

The reason the crescent moon has been given this honor has to do with the reciprocity of phases between our two worlds. If you lived on the moon you'd be very aware of the beautiful Earth dominating the night sky, with ever-changing cloud formations and phases. The light from Earth, four times the diameter plus at least *five times shinier* than the moon appears in our own sky, bathes the lunar terrain with near-blinding brilliance. This intensity varies, of course, with Earth's phase, which exhibits shapes that complement the moon's. The slimmer the moon looks from here, the fatter and brighter our planet appears from there.

You can also think of earthshine as sunlight taking a triple voyage. When seeing it, you're viewing sunshine that has first struck our own world, bounced to the moon, and then bounced back to your eyes. Because of this extra travel time, earthshine is "older" than light from the brighter sunlit crescent. If the sun should blow up and suddenly go dark,

Earth barely moves in the lunar sky. To see Earth near the moon's horizon, one must move to the poles or to the east or west limb. Here we look from the rim of Orientale Basin.

the crescent moon would vanish at the same time. But the moon's earthshine would continue to glow for a few additional seconds. If you see this happening, grab the phone and sell your stocks.

As the moon grows (waxes) each evening into a fatter crescent, its points or horns always aim leftward. Each night it jumps farther from the sun by twenty-six times its own diameter and increasingly becomes a resident of a dark, rather than twilight, sky. Simultaneously the earthlit portion dims, since Earth is shrinking into a thinner phase in the lunar sky. By the time the moon is four days old, it's separated from the sun by more than a 45-degree angle and doesn't follow the sun into the ground until after nightfall. From this day forth it's a denizen of the night rather than twilight.

Now that it's easier to see, we can leisurely consider its size. The famous "moon illusion" always makes the moon seem much bigger when it's low near the horizon, and smaller when it's higher (see page 270). The effect is so vivid that artists routinely get away with depicting the crescent at least fifty times bigger than it actually is.

As the figure on page 144 shows, appearance and reality are opposites when it comes to the rising moon. When the moon is low, that's just when

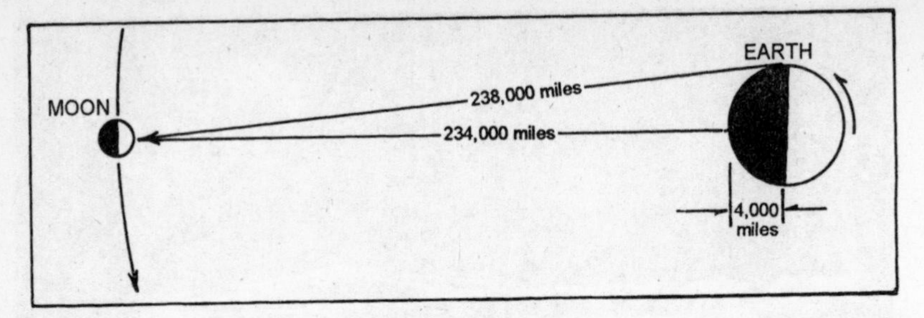

When near the horizon, the moon is 4,000 miles farther away than it is when at its highest point in the sky. So, contrary to the famous moon illusion, the moon is smaller when rising.

it's farthest from our eyes and smallest. When it's high up, our spinning world has carried us over Earth's hump and closer to it by as much as 4,000 miles, making it almost 2 percent larger. You'd confirm this if you measured the moon's size in the sky—yet there's no denying that the opposite seems true, making the moon illusion one of the most powerful and persistent in the heavens. Even to people who fully understand the effect's imaginary nature, the low moon continues to look enormous.

More peculiar than this, and harder to understand, is the way the moon's size remains exaggerated no matter where in the sky it appears. Having observed it throughout your life, try this test to see how you perceive the moon's dimensions:

If full moons could be piled one atop another, how many would be needed to stretch from the horizon to directly overhead?

Choose one answer: (A) 20; (B) 40; (C) 75; (D) 100; (E) more than 150.

Most people choose B. Try it on others and you'll find that virtually everybody remembers the moon as much larger than it really is. In reality, the moon's so small in our sky that a pile of 180 is needed to reach from horizon to zenith: choice E. It's so tiny that if you wanted to pack the entire sky with moons, you'd need 105,050 of them.

To appreciate the moon directly, face-to-face, as our ancestors did for numberless millennia, we have the incomparable advantage of viewing it through inexpensive but spectacularly effective instrumentation. The ideal time for such direct observation is when the crescent phase metamorphoses into the half-moon. It's best, then, to hold a moon-observing party between the fifth and tenth days after the new moon—a time when it's nothing short of awesome through any instrument and with any magnification. In this part of its cycle it will also conveniently be up as soon as darkness falls—count on it.

We now skip through the full phase around day 15, when the moon (contrary to what the average person believes) is least interesting telescopically (see page 52). On the twenty-third day the crescent reappears on the side of its orbit that doesn't allow it to rise until after midnight. This monthly five-day appearance of the waning crescent is thus a privileged sight for insomniacs and those on the graveyard shift. Its horns (points) aim to the right, and its thinness tells you the time of night. A fairly slim crescent rises just ahead of morning twilight; an extremely thin crescent is seen only *in* morning twilight, and is low in the sky to boot.

Because the waning crescent incarnates during the night's second half, the week of its existence confers darkness in the hours before midnight. The period between last quarter and a few days after new moon is thus the dominion, for all practical purposes, of moonless skies. That's when we unwrap the celestial gifts whose full glory demands the black of night.

Then the *rest* of the universe awaits.

Unidentified Flying Planet

The night's brightest object is, of course, the moon. But what's second brightest?

You don't know? Join the crowd.

The real black hole enveloping our world is the extensive lack of elementary sky knowledge. That's why a dazzling apparition can dominate the sky, generate UFO reports, shine a hundred times more brightly than the brightest stars, and yet travel incognito. It's Venus—popularly called the Evening Star.

Every nineteen months, when the Evening Star reaches its maximum brilliance and lights up the western twilight like a searchlight, Earth's inhabitants go on a binge of misidentification. Some years, according to one prominent UFO author and researcher, bright planets (with Venus leading the charge) account for more than half of all UFO reports, and such sightings don't all come from dimwits. Jimmy Carter, while governor of Georgia, phoned the state police to report a UFO that proved to be Venus. A few years later the *CBS Evening News* featured footage of a UFO filmed by an Australian camera crew; this UFO also turned out to be the cloud-shrouded planet. And a squadron of Allied bombers returning from a mission over Japan in World War II saw a brilliant light that appeared to keep pace with them. Firing their guns, they attempted, without success, to blow up the Evening Star.

Passing clouds sometimes create the illusion that Venus is in motion. Its creamy-white untwinkling radiance, oddly steady for such a low object, is bright enough to cast shadows on white surfaces when seen from an unpolluted site. It is the night's most riveting object. Even shadows cast by its light are noticeable.

People don't think about shadows very much, but I remember as a child being bewildered by their changing properties. It seemed baffling that shadows were crisp and sharply outlined when falling on a nearby object, but blurry-edged when more distant. Check it out: In sunlight, put your hand a few inches above a piece of paper. The shadow is stark and crisp. But look at the shadow of a tree or building on the road or sidewalk, and the outline is fuzzy.

The changing properties of shadows bugged me enormously, yet I never mentioned it to anyone. (How many such hangups slip through the cracks in the psychological evaluations of kids at school? All it would take is one extra question—"Do shadows sometimes bother you?"—and schools would be able to help kids like me.) Answer: The source of light, the sun, is not a point but an extended object. It has size. A building's shadow has a blurry edge because it's the transition zone between no sun and full sun, the place where the sun is fractionally obstructed. An ant walking across the shadow boundary would see the sun *partially* eclipsed!

That's why those tiny brilliant halogen reading lamps create such stark illumination. Since the bulb's filament is tiny, all shadows are abrupt. Shadow boundaries produced by ordinary light bulbs are much fuzzier, while fluorescent bulbs are so long they come close to producing no shadows at all!

It means that a trip to, say, Pluto, where the sun is an intense starlike point, would bring spectacularly crisp shadows everywhere, the kind of alien experience we'd *want* to encounter on a faraway world. Here on Earth, sharp natural shadows are confronted only during the moments before and after the totality of a solar eclipse, when the sun is reduced to a pinpoint of light, a phenomenon occurring just once every few centuries for any given location.

Sunrise is very slow on Mercury. The sun takes eighty-eight days to go from horizon to horizon, and appears 3.3 times larger than from Earth, producing strangely indistinct shadows.

Such strange shadows are experienced just one other way—by the light of Venus. I saw such razor-sharp shadows just once, in the sand of a remote South Pacific island. I had expected Venusian shadows to be dim, but was unprepared for their razor-sharp edges, the natural result of Venus' being a point source.

Going the other way, few astronomical artists correctly paint scenes on Mercury with wildly blurry shadows, which would occur because of the sun's huge size as seen from that innermost planet.

(Mercury's a strange land for other reasons as well. Its day-night temperature variation is 1,000 degrees Fahrenheit, far more than that of any other planet. Daytime sees such fiery heat that lead, tin, and even zinc would melt into puddles. Yet protected valleys at the poles are simultaneously covered with ice!)

It's also seldom realized that the moon's shadows on a snowy landscape have properties identical to the sun's, since both disks appear the same size.

As for Earth's shadow in space, it is both fuzzy and *red,* the result of sunlight's being bent upon passage through our encircling atmosphere. Air throws refracted light from all the world's sunrises and sunsets into our shadow, which explains why a total lunar eclipse—when the moon flies into Earth's shadow—finds the moon assuming a strange, coppery tint.

That Venus is the only starlike object to cast shadows demonstrates its great brilliance, dozens of times greater than the night's brightest star, Sirius. This kind of radiance, which makes Venus so spectacular to the

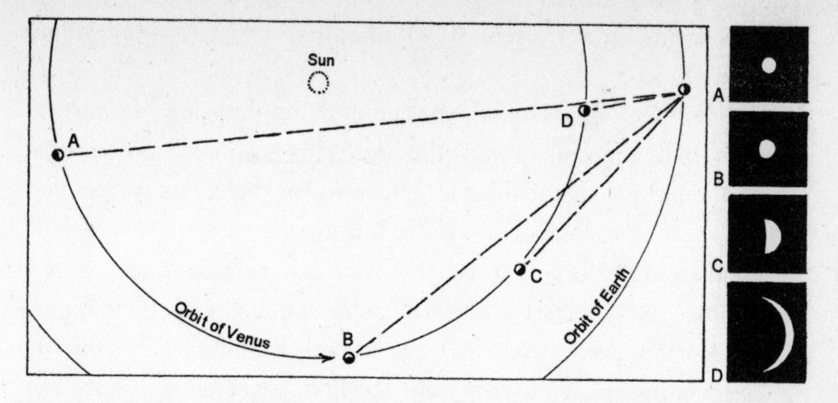

Venus, moving faster than Earth, passes us every 584 days. At its farthest, Venus is 160 million miles distant, but only 26 million miles away at its nearest approach. The images at right show how the planet appears in a small telescope at the positions shown.

naked eye, works against us when observing it through telescopes. Against a dark sky such blazing contrast is more than the eye can handle, and its form becomes smudged. The best solution is to watch Venus in twilight or even daylight, when it's higher up as well. But even in twilight it's impressive telescopically only when close to Earth and illuminated from behind by the sun; then it forms an intriguing crescent large enough to look stunning in even the smallest telescope.

Creamy white, and as devoid of detail as a political speech, the crescent of Venus can swell to one-tenth the apparent diameter of the moon. At such times (see table, page 152) it's prominent through plain binoculars. There are even reports of people's resolving the Venusian crescent with just the naked eye. Telescope users can follow its changing phases, and perhaps glimpse the curious **ashen light,** an enveloping halo that spills around the entire planet to produce a complete ring when the crescent is very thin. It's caused by the planet's thick, noxious atmosphere of carbon dioxide and sulfuric acid droplets.

Visits by numerous spacecraft have revealed Venus to be the most unpleasant planet in the known universe. Purgatory? No, much worse: It's the Other Place, the destination of people who do bad things like touch lenses with their fingers. Temperatures on its surface stay uniformly hellish, as if regulated by a thermostat from some stygian sauna. Day and night, the forecast never varies; a Venusian meteorologist could play the same recording forever: overcast, with highs around 850 degrees, lows around 850.

Venus is much more scorching than even sun-hugging Mercury because its thick atmosphere traps the heat. This is a model **greenhouse effect,** the place that made that term famous long before its current popularity as applied to Earth's own global heating.

Sounds terrible? It gets worse. The Venusian surface is subjected to ninety times the air pressure of Earth, achieving the most efficient pressure cooker in the solar system. A few seconds would turn the cow who jumped over the moon into beef stew. The first spacecraft, a Russian *Venera* in 1970, lasted just twenty-three minutes before it could no longer transmit its nightmarish report from the surface.

Though Venus is sometimes called our "sister planet" since its diameter and density are nearly the same as ours, all family resemblance ends right there. Goddess of love, certainly—as the night's brightest "star" it's appropriate that Venus be forever associated with love, a link strengthened by the infamous 1959 Frankie Avalon song. But it's strictly a "look but don't touch" affair.

Yet no unpleasantness greets our senses when we observe the Evening Star reigning over the fading twilight. Quite the contrary, its apparition exudes a strange serenity, a timeless presence. Clearly it's always been that way; Venus' appearance in art and poetry has spanned all centuries and cultures.

Venus is the only "star" to appear conspicuously in the daytime to the naked eye. To see it easily after dawn, use the table to determine when Venus is prominent as the Morning Star, rising ahead of the sun. Then keep your eye on it as twilight brightens, and you'll find it lingers even after sunrise.

Or try for it against a blue sky in late afternoon during the times when it's brightest as the Evening Star. Keep your eyes in shadow so that your pupils do not constrict. Huddle near a building's edge to block the sun while scanning the sky. And hope that, if a policeman passes by, he believes your story.

How many other celestial bodies can make the same claim of daytime visibility? Jupiter, certainly. And reliable reports indicate the Dog Star, Sirius, is marginally within view. Anything else is dubious. For almost two centuries, many have claimed that numerous stars appear by day when observed from the bottom of a chimney, mine shaft, or similar tunnel. I am periodically asked about this controversial phenomenon, always by older people, indicating that a great deal of popular publicity about it occurred earlier in the twentieth century. Modern investigators, however, call this an old wives' tale.

If, for some peculiar reason, you feel like helping to settle the argument, you'll encounter few problems if you're one of the tens of millions who live between latitudes 38 and 40 north. This zone includes a dozen major cities: Philadelphia, Baltimore, Kansas City, Indianapolis, Denver, and San Francisco, to name a few. From this region, the brilliant star Vega passes directly overhead daily. While Vega blazes at its best on summer nights, this experiment seeks its appearance as a *day star;* a **planesphere** (a wheellike device displaying the orientation of the stars on any chosen date and time) can reveal the dates and times when Vega is straight up. One consistent occasion, for example, is 4 P.M. on Halloween day. At such a time, see if you can spot it by looking up a chimney. For the small price of being covered with soot, you might see a faint dot in the sky. And settle a controversy.

There is no doubt, however, about Venus' blue-sky visibility. The only challenge is finding the right spot at which to look. When our sister planet is near its brightest (see table, page 152), just mark off about three extended fists held at arm's length to the upper left of the setting sun—and there you will find it!

Because the highway of the moon and planets—the zodiac—regularly sweeps those bodies past Venus, we're periodically treated to dazzling con-

A global view of the surface of Venus, captured by the cloud-piercing radar of the Magellan *spacecraft in 1991.*

junctions involving the Morning or Evening Star. The most frequent and spectacular Venusian patterns feature a close rendezvous with the moon, which is always a crescent for such beautiful twilight encounters.

Venus at Its Best

The dates show Venus' *greatest elongations,* when that planet and the sun are maximally separated in our sky—by a 45-degree angle. **M = morning sky;** look east before sunrise. Venus is usually prominent from six weeks before to sixteen weeks after these dates. Binocular users: Best views are six to nine weeks *before* these dates. **E = evening sky;** look west after sunset. Venus is prominent from two to four months before to six weeks after these dates. Binocular users: Best views are six to nine weeks *after* these dates.

1995: Jan. 13 M	2001: Jan. 17 E
1996: Apr. 1 E*	2001: June 8 M
1996: Aug. 20 M*	2002: Aug. 22 E
1997: Nov. 6 E	2003: Jan. 11 M
1998: Mar. 27 M	2004: Mar. 29 E*
1999: June 11 E*	2004: Aug. 17 M*
1999: Oct. 30 M*	2005: Nov. 3 E

Venus' greatest brilliancy occurs a month after evening elongations (E) and a month before morning elongations (M).

*Highest-up, ideal apparitions of Venus.

All Venusian apparitions are not equal. Because of Earth's tilt, Venus is much more prominent for Northern Hemisphere observers when it reaches greatest evening elongation in the spring, and when greatest morning elongation occurs in the autumn. Interestingly, both such favorable appearances usually occur in the same year, as in 1996 and again in 1999; these are unusually superb years for both the Evening and Morning stars.

Sapas Mons is a three-mile-high extinct volcano sitting near the equator of Venus. This 1992 view by the Magellan *spacecraft exaggerates the height of features tenfold, and displays a black sky. The actual, overcast sky above Venus is gray.*

You, on your own, can easily determine future apparitions of Venus. In the same period that Earth orbits the sun eight times, Venus circles thirteen times. That is to say, every eight years Venus returns to its starting position, from our perspective. It then exhibits the same brightness, appears in the same constellation, at the same height above the horizon. In the table, notice how Venus is maximally separated from the sun's glare and thus at its best in January of 1995. We simply add eight years and there it is again, in January 2003.

As another example, the pair of particularly favorable appearances by Venus in June and October 1999 will be repeated during those same months in 2007.

A truly awesome if esoteric experience involving Venus is to let it define the plane of the solar system, which bestows a new perspective to everyday reality. Here's how:

First, you acknowledge candidly, as if at an AA meeting, that you've always used the *horizon* as the main reference point when viewing the sky.

Locate Venus just after the sun sets . . .

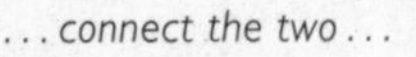

. . . connect the two . . .

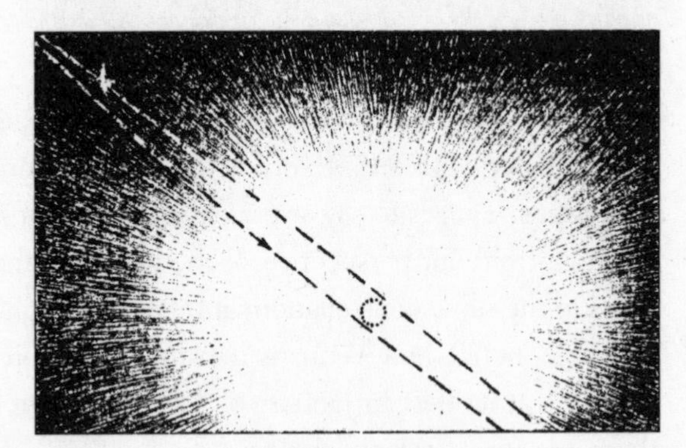

. . . then disregard the horizon, and you're left with a view of the solar system's plane.

That line separating earth from sky has always seemed important, while other references were mere abstractions. For instance, only in textbooks did illustrations of planetary orbits make sense; in everyday life, planetary orbits are nowhere to be seen. The point is, they *can* be real, they *can* be seen. Venus allows us to do this.

Shortly after sunset we note the brightest spot of twilight, which shows us where the sun is lurking below the horizon. Then we draw a mental line from the sun's position to Venus, and we realize that *this* marks Venus' orbit in space. And since all planets orbit the sun in approximately the same plane, we are now viewing the plane of the entire solar system! We then quietly let go of the horizon; what remains is a presentation of the actual solar system, revealed in its grandeur for the first time. Suddenly a greater perspective, a deeper point of reference emerges in the heavens. It is a wondrous experience.

You can consult the table and deliberately search for Venus. But the more venerable method is simply to stumble upon the Evening Star by accident while taking a twilight stroll. Your gaze, sooner or later, will automatically be yanked upward by its brilliance. Venus is so eye-catching it's unnecessary to seek it out.

On its own, it will find you.

Tales of the Dipper

Many of us regard the Big Dipper as an old childhood friend, and can remember the night some farsighted grownup first pointed out its seven stars. Its familiar shape will grow increasingly distorted with the passage of millennia: The five middle stars are heading in one direction while the other two drift a different way. But to our eyes it hasn't changed a bit since we were kids.

Poetically, spring itself, the season of renewal, is when the Dipper reaches its annual apex. These nights it hovers high in the north, almost overhead, a far cry from the horizon-skimming stance it adopts in the autumn.

While the southwest now regales us with an upbeat display of dazzling stars, it's a gothic story when we look northward, where the Dipper floats forlornly in a dark and desolate region of the sky. This realm lies far from the Milky Way, far from the plane of our own galaxy. The Dipper therefore guides our eyes away from the city lights of our own celestial neighborhood toward the capacious richness of the rest of the cosmos.

We can think of our solar system's existence as a speck imbedded within a pancake of half a trillion glowing grains. When we look along the thick part of the disk (as we do when facing southward in late summer) all

From Big Mess to Big Dipper —and Back Again

The Big Dipper changes shape in time. Alpha (α) and eta (η) are moving contrary to the other five stars.

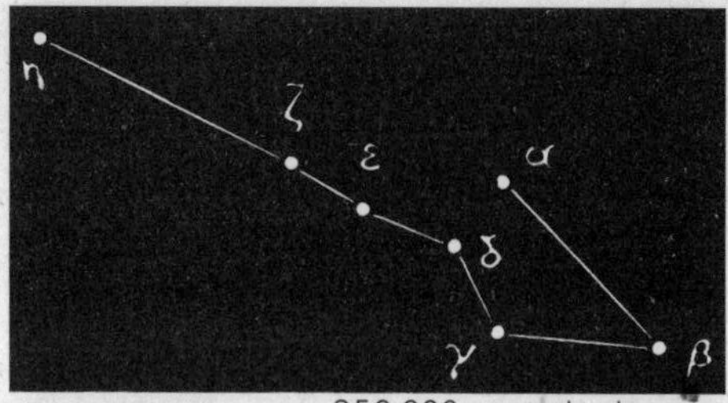

250,000 years in the past

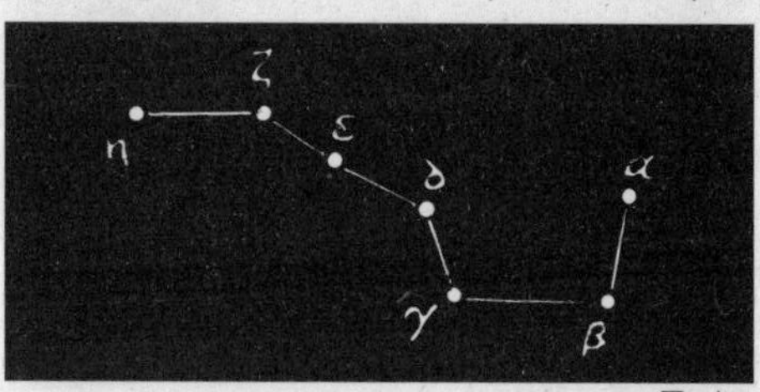

Today

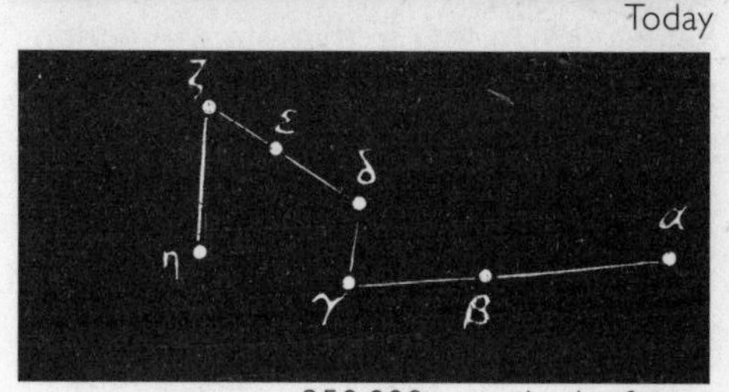

250,000 years in the future

we see is material inside our own galaxy: countless stars, foreground clouds of gas and dust, and the creamy faraway glow of endless unresolved suns—the soft radiance of the Milky Way. But when we gaze toward Virgo or the Big Dipper we're looking up through the thinnest part of the pancake. Few foreground stars and no dust obscure the view, so that direction offers a crystal-clear window straight out of the Galaxy to the yawning chasm of space.

In this region, even amateur telescopes unveil an extragalactic potpourri that includes the strange, violent galaxy M82 and the awesome spirals M101 and M81. Each has more stars than could be counted nonstop since the telescope's invention—and the distances between each of them and us almost defy comprehension. By the time their incoming images pass the Dipper's stars, the light from these galaxies has already completed 99.999 percent of the journey to our eyes, with only 100 more light-years to go.

The beautiful galaxies M81 and M82 lie about 10 million light-years beyond the stars of the Dipper. If you could travel fast enough to get from New York to Tokyo in one twentieth of a second, you'd need 10 million years to reach these cities of suns. Yet they are among our closest galactic

M81 is one of the nearest and most beautiful spiral galaxies. It appears among the stars of the Big Dipper, but is actually one hundred thousand times more distant.

neighbors, hovering nearer than all but a hundredth of 1 percent of the 10 billion galaxies detectable through today's telescopes.

Compare them, for instance, to the stupendous Ursa Major Group lying much farther in that same direction. *That* cluster of galaxies is so remote that the glow of its quadrillion suns, departing just as the last dinosaurs gazed upward, will not reach us for another 700 million years! Its ancient light, flushed like an old sepia print and altered by the expansion of the universe, brings us the latest news about events and objects that no longer exist.

All this ancient and distant beauty lies at staggered and staggering distances beyond the sparse, lonely stars of the Dipper. The Big Dipper's dark and quiet corner of sky is thus an illusion, a mere marker, a buoy in

M82 appears to be a spiral galaxy in the process of exploding. In fact it is producing billions of new stars out of a huge cloud with which it is colliding. At 10.5 million light-years' distance, it's the nearest "peculiar" galaxy.

M101 can be found just above the end of the Dipper's handle. The star nurseries in its outer spiral arms are so bright (in most telescopes) that they are listed with their own numbers in the New Galactic Catalogue.

the night's ocean to show us where we must turn our instruments to plumb the treasures looming in its depths.

Curiously, the Big Dipper is not even a constellation in its own right. It's an **asterism,** a *segment* of Ursa Major, the Big Bear. *Big* is the correct adjective; it's the third-largest constellation in the heavens. It barely missed first place, since there's a virtual three-way tie for Largest Star Pattern. Virgo voluptuously spreads herself across only a few more degrees of sky, as does Hydra the Serpent. On the other hand, the Southern Cross, the smallest constellation, could fit within the Bear's boundaries nearly twenty times over.

It's odd, to say the least, that so many ancient civilizations discerned the shape of a bear in this region of sky. True, when one has a few too many drinks, old Smoky might appear minus his hat. But a more rational pattern is the "Plough" seen by Britishers or the "Dipper" seen by Americans. A bear is stretching it, and yet that is exactly what Native Americans, ancient Greeks, the Germanic tribes of middle Europe, and others saw in this formation.

Why such disparate civilizations should all project the same unlikely bruin onto these northern stars remains a mystery. In any event, such a venerable tradition inspires today's stargazers to follow their lead and mentally trace out a bear in the stars of Ursa Major, even though this perpetuates a mild form of collective delusion.

In the ever enjoyable one-upmanship department, the kid in each of us can impress just about anyone by rattling off the names of the Dipper's

Fitting an anatomically accurate bear to the stars of Ursa Major isn't easy. Renaissance Europeans cheated by giving the bear a tail as long as a lion's (inset).

stars. This works because, while the Dipper is so universally familiar, few have bothered to make friends with its individual stellar members. There are no more of them than Snow White's dwarfs, so it's not much of a task even though a few of their names do sound a bit Dopey to Western ears.

Dubhe (say: dubby) for instance, makes children giggle, while Merak (mee-rack) seems like a good name for an alien visitor. Yet this pair, lying at the leftmost edge of the Dipper's bowl these spring nights, are among the most famous stars of the heavens—they're the famous "Pointers" that guide the celestial beginner to the North Star. A line connecting the two and extended downward takes our eyes to Polaris, the polestar for the next millennium.

If we pretend Polaris and the Pointers form an hour hand of a clock, it will aim straight up and announce the correct time at midnight during the first half of March. An enormous nocturnal Big Ben, it's an inexpensive way to tell the time, even if it works for only a fortnight each year. (You get what you pay for.)

Continuing around the Dipper in order, we pass Megrez, the faintest of the seven stars, reach Phecda at the junction of bowl and handle, and then the three handle stars: Alioth, Mizar, and Alkaid at the tip.

Notice that the commonest beginning for these stars is *Al*. It doesn't derive from *Alexander, Alan,* or *Albert*. All the Dipper's names come to us from the Arabic, and the favoritism toward *Al* is because *Al* means "the"

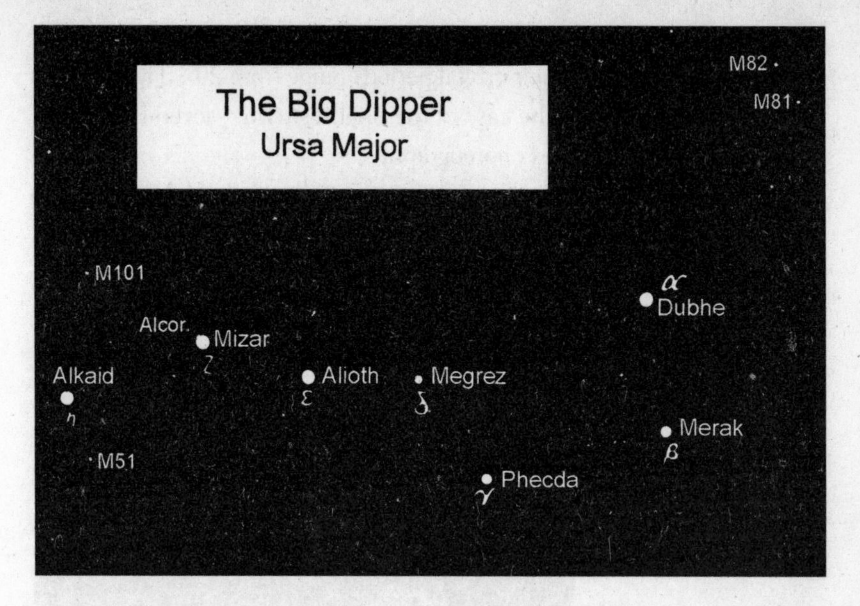

in that language. *Alkaid,* for instance, translates as "the leader" (despite its being the trailing star in the Dipper's nightly rounds).

Readers familiar with the night's stars can think of many other famous "Als" scattered around the sky, like Aldebaran, Algol, Altair, Alphard, and so on, all of which originally meant "the this" or "the that."

Another Dipper *Al* is Alcor ("the abandoned one"), faintly seen right next to Mizar, the handle's second star. This pair constitute the most famous double star in all the heavens, and are still sometimes called by their ancient title of "horse and rider." The ancient Arabs often tested their sight with this duo, believing the ability to see Alcor indicated keen vision. Indeed, fourteenth-century Arabic writings sometimes referred to this star as the Riddle.

Today that title is as appropriate as ever—for a different reason. Either "the abandoned one" has brightened over the centuries or the ancient desert dwellers had vision problems, because Alcor is obvious today even in polluted city skies. The riddle is how anyone can *fail* to see it.

Point a telescope its way and a third star pops into view right next to Mizar. All are physically related rather than merely lying in the same line of sight. This, then, is a triple-sun system, and the very first multiple star to be photographed (back in 1857).

Spectroscopic studies have shown that Mizar has a couple of unseen suns circling it as well, making it a five-star system, three of whose members lie too close together for telescopic detection. Dancing around each other in complex paths, their orbital periods range from 20½ days to half a year to four years to—in the case of the brightest two—several thousand years, a grand and complex choreography.

In fact, most of the Dipper's stars are gravitationally linked. Not simply lying in random line-of-sight alignments the way nearly all other constellations are, they're family members forming a large and unconfined cluster, or **association.** Of the thousands of such star groups known, the Dipper is the nearest to Earth, which is why it appears so large. The distance to this Big Bear Brotherhood is about 80 light-years, nearby as such things go but still a formidable expanse to travel. Even if the vehicle you're hitchiking on is the space shuttle, a journey to the Bear would take 2 million years.

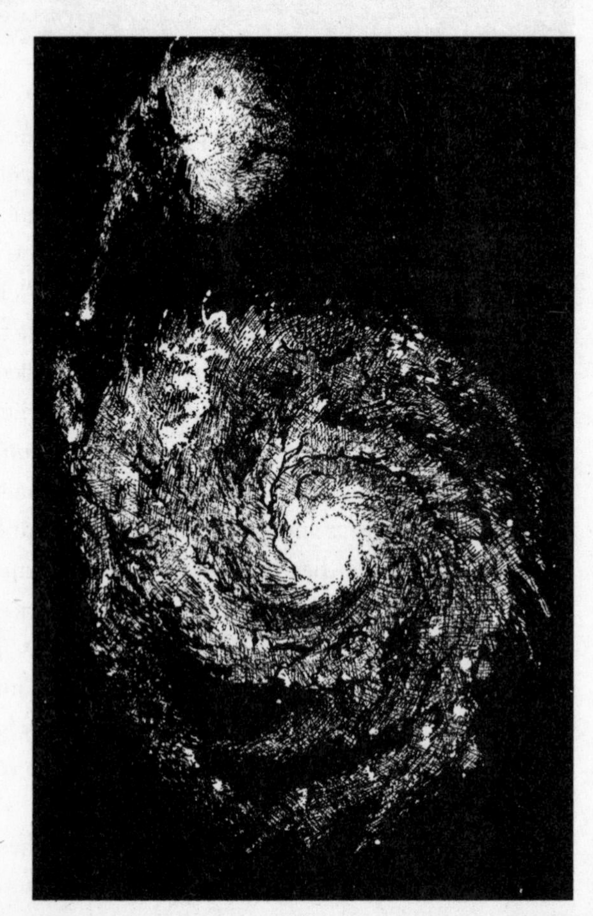

M51, the Whirlpool Galaxy, is a spiral just under the end of the Big Dipper's handle. The smaller galaxy NGC 5195 has brushed past M51, tidally disrupting the spiral arms. In 1994 a supernova was seen at the center of the Whirlpool. But it really exploded 35 million years ago.

It's also possible to cheat, big-time, with the Dipper stars and rattle off their "Bayer system" names, if you know the first few letters of the Greek alphabet (see table, page 4). Dubhe is Alpha, and the other letters of the Greek alphabet simply sweep around the constellation in order, without regard to their brightness. So you can make quite an impression by rattling off Alpha, Beta, Gamma, Delta, Epsilon, Zeta, and Eta Ursae Majoris (say: UR-see ma-JOR-is) and simply not bothering with their ancient Arabic names. Anyway, "Dubby" sounds ridiculous. Call it Alpha Ursae Majoris and nobody will laugh.

Don't want to bother with names and facts? It's more than enough merely to observe the Dipper the way you did that night long ago when you first saw it. Neither rising nor setting but wheeling around the North Star without end, such a dependable nocturnal guide is worthy of a springtime salute.

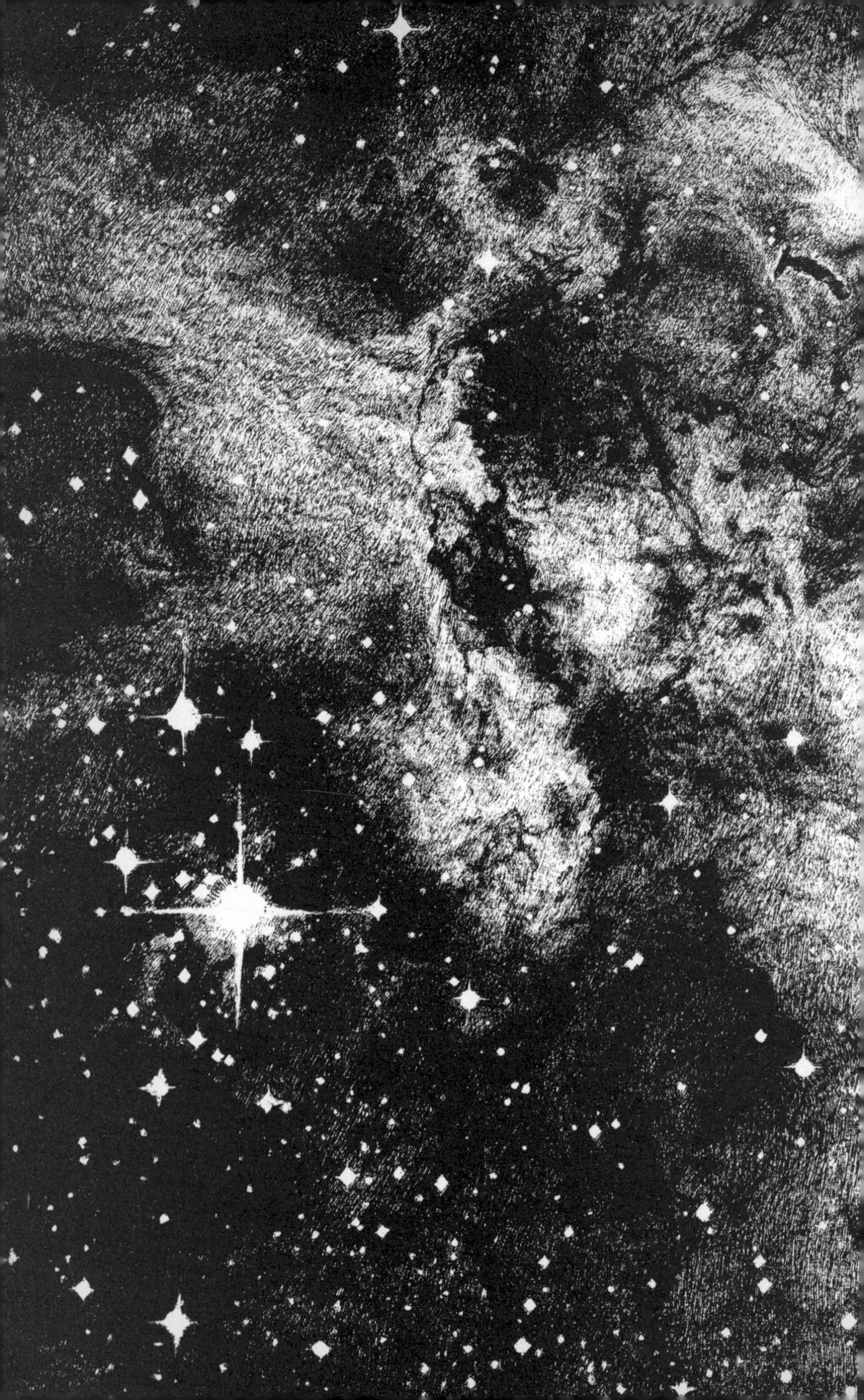

Summer

Nowhere in our galaxy have we found a star that burns more intensely than Eta Carina (center left). *It is 3 million times as luminous as our sun, and the violent solar winds of Eta and its massive neighbors have blown the surrounding gas and dust into a contorted nebula.*

Inside a Black Hole

The strangest thing in the universe?

Nobody's taken a survey, but a black hole would probably top the list. First envisioned in 1798, black holes weren't taken seriously until a few decades ago. Now they're an obsession, epitomizing mystery and danger like no other object ever known. They're the only things throughout time and space whose essence is utterly inexplicable by science, prowlers in the inaccessible alleyways beyond our comprehension.

There's also this little matter: Nobody's ever seen one or ever will. There's no certainty they even exist. Yet these invisible maybes, the most cryptic places in the cosmos, could be conduits to other dimensions that hold the key to the future of the universe itself.

We can ease our way into this shaky realm by looking straight up at midnight in midsummer, when the surest black hole floats directly above us. At that hour the Milky Way splits the sky neatly in half, the lovely constellation Cygnus the Swan flying along its creamy glow. One moon's-width east of the star Eta, which marks the celestial swan's neck, lies a blue star just bright enough to appear in binoculars. Going by the catchy name HDE226868, it's about as heavy as a dozen suns.

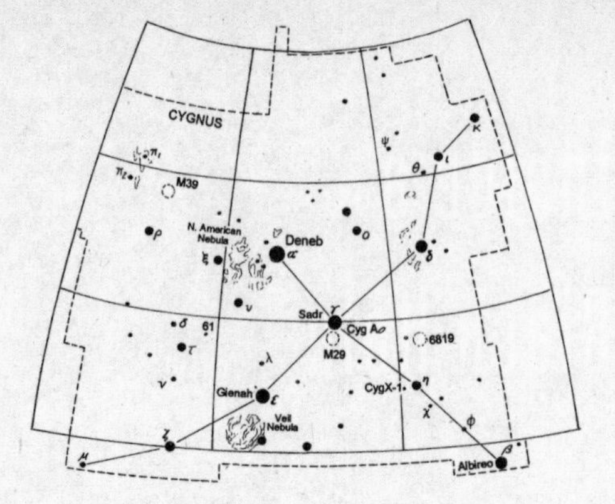

Cygnus X-1, a likely black hole, is overhead on summer nights.

Something very fishy is going on here. First off, this bloated blue sun whirls in a circle every 5.6 days, as if caught in the gravitational grasp of an immense object. Yet the celestial sumo wrestler that is twirling HDE226868 around like a puppet is strangely invisible. Massive stars are always very bright; this star-hurler should be brilliant. Instead, the most powerful telescopes reveal no trace of anything there at all. Thus we have exhibit A: a heavy, underluminous object.

Next, this spot of sky emits an intense beam of X rays, a rare and powerful energy that's always a sign of violence. Physics tells us that anything spiraling toward a black hole should be whipped to such frenzied speeds that X rays are thrown off; sure enough, you can practically see your bones when you raise your hands in its direction. Could this awesome X-ray factory be fed by material from the visible blue star as it falls, helpless, into a black hole's grasp? The mysterious X-ray beacon is such an important clue that this likely black hole is usually known simply by its name in X-ray catalogs: Cygnus X-1.

There's more: Tremendous changes in the X-ray intensity occur in less than five hundredths of a second, an eyeblink. Such snappy variations prove it's a tiny object. Why? Think about it: If our own sun tried to flicker, or vary its light, and somehow managed to alter its entire huge body all at once, the changes would still require several seconds to take effect because light from its surface closest to us would arrive sooner than the energy from its edges—since light takes five seconds to cross the sun's disk. Cygnus X-1's instantaneous changes prove it's no larger than 100

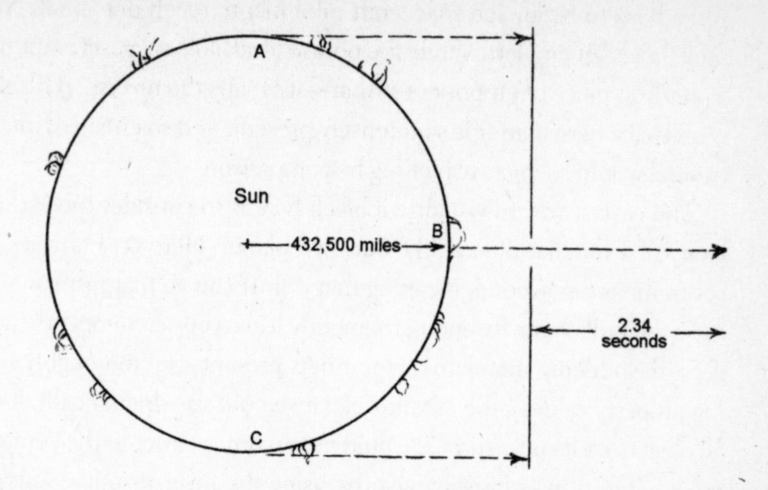

Light leaves the sun simultaneously from A, B, and C, but rays from the middle of the sun (B) get a 2⅓-second head start.

miles across—one-twentieth the size of the moon! This raises another marvel: How could something so small send such a powerful beam of energy to our satellite-borne X-ray detectors?

Put all the evidence together and you've got a strong case for a black hole. Would it stand up in court? Probably not; it's circumstantial. The best we can say is that a black hole explains everything we see in this bizarre district of the sky. If Cygnus X-1 is *not* a black hole, then we have absolutely no idea what this tiny, heavy, violent, shadowy entity can possibly be.

The dimension into which black holes take us is more bewildering than a Martian road map. But many features of these mind-stretching objects are simple; it's no great task to understand how they form, what they are, and whether they pose a danger to us. We should resist the temptation, however, to lump them into a single category as if all were built at the same plant; black holes need not be confined to star-sized units. Microscopic versions may pepper the universe, and colossal black holes weighing as much as 100 million suns probably sit at the centers of many galaxies.

Black holes, undeniably, have had bad press. Their poor reputation suggests people distrust them, suspecting they'll gobble up the rest of the universe if given the chance. In reality, black holes are usually so small you'd have to be an ace spacecraft pilot just to reach one at all. Nor are they "holes" of any sort. While the phrase *black hole* suggests emptiness—a poorly lit piece of emptiness at that—it's really the reverse. A black hole is a place where matter is so intensely present, and so crushed, that a bar of steel would seem a wisp of fog by comparison.

The easiest way to visualize a black hole is to consider the escape velocity of a familiar object, like our own planet. Here on Earth it's about seven miles per second, meaning that only if you go that fast can you escape the pull of gravity and permanently leave your creditors behind. But if Earth suddenly shrank to one-tenth its present size (thoroughly confusing property values), the escape velocity would rise dramatically, because all objects on its new surface would be ten times closer to the center than before. Ten times closer means (by using the inverse-square law, which operates throughout the universe) that everything would be *10 squared, or 100 times* heavier than previously, requiring a much higher speed to fly away from this new shrunken Earth. If Earth could be compressed smaller and smaller, there'd come a point where the velocity needed for escape would equal the speed of light, 186,282 miles per second. Then you could love it but not leave it: If light, the universe's fastest thing, could not exit, neither could anything else.

A black hole could be compared to a bathroom sink.

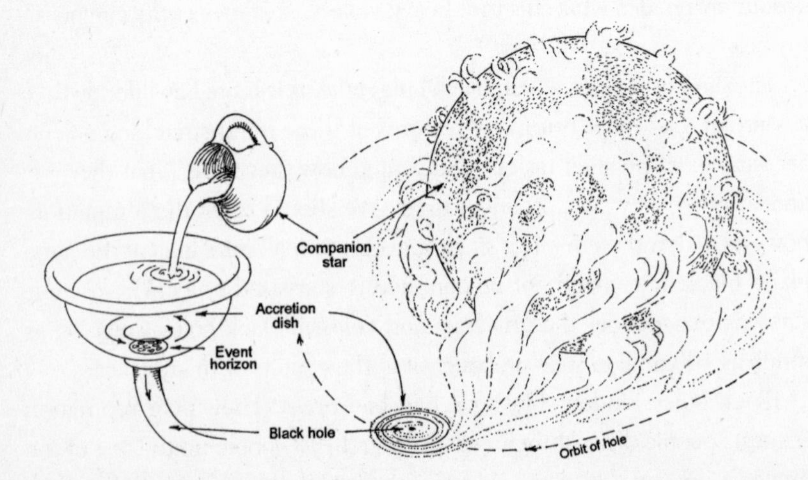

Any object could thus become a black hole if squeezed enough. To reach black hole density, Earth would have to be crushed down to the size of a marble. Can you imagine a marble weighing 6,000,000,000,-000,000,000,000,000 tons? The point is, all it takes to be a black hole is sufficient compactness for gravity to get strong enough so that a speed greater than light is needed to escape. Mount Everest, under those conditions, would become a black hole if every boulder and truckload of its material were crammed into a ball the size of an atomic nucleus.

Black holes are scarce because matter normally does not voluntarily pack itself so firmly. Among the many forces that resist the process is heat—meaning atomic motion—which doesn't like to be cramped. Then, too, the outer shells of atoms are electrons, all with negative charges, and as we've learned from childhood, like charges repel, while opposites attract. So atoms naturally give each other space. On an even smaller scale, there is "degeneracy" pressure; even if you break atoms, their inner electrons need some breathing room and resist being crammed together. All of this helps prevent things in the universe from getting too densely packed, a process that somehow fails in the New York subway system.

It would seem that a black hole manufacturer would have an insuperable problem turning out samples. No machine could possibly crush and cram things so tightly, and surely matter would never do the job on its own. How then can black holes possibly form?

The simplest mechanism involves obese stars—those more than 3½ times heavier than the sun—going through late-life crises. Such stars are relatively scarce, less than 1 percent of the stellar population. But with half a trillion stars just within our own galaxy, there are still plenty of candidates.

What happens is that in a star's old age (when the nuclear furnace in its core loses its punch and no longer emits enough outward-pushing energy) it cannot resist the gravitational urge to collapse. Stars, after all, are gas, and gas is easily compressible, like the air in the tanks used by scuba divers. When the star starts contracting, its entire surface is suddenly closer to the center than before, and the inverse-square law pops into play again. Every particle on that surface now grows heavier, causing the star to collapse even more, which makes it heavier still. And on it goes, a runaway shrinkage like some dry cleaner's nightmare. When the star becomes

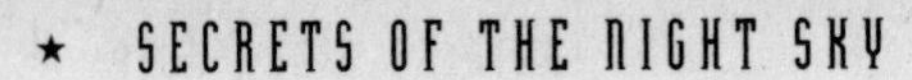

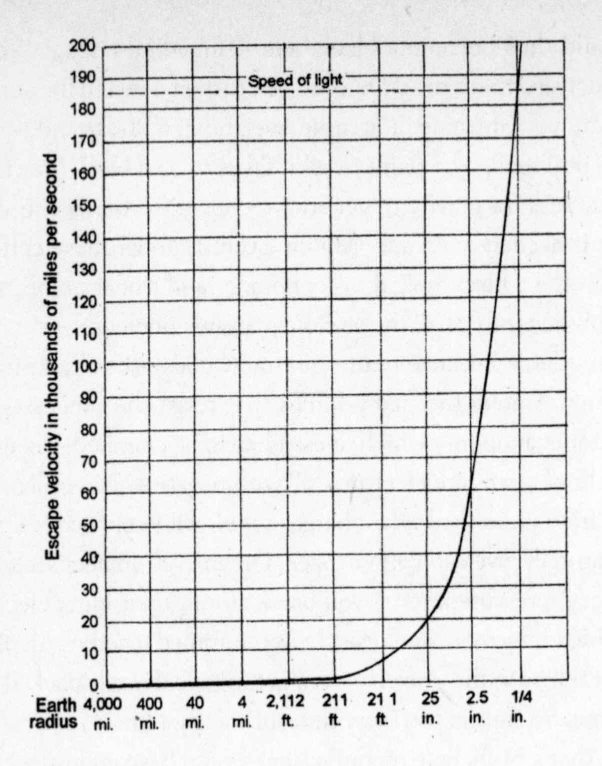

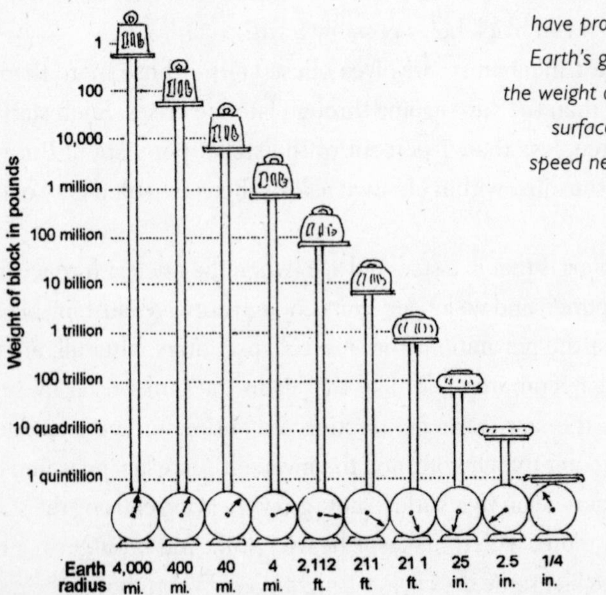

If Earth shrank, its density would increase. This would have profound effects on Earth's gravity, increasing the weight of objects on the surface, and hence the speed needed to escape.

three times smaller, material on its surface weighs nine times what it did originally. When the star is five times smaller, the weight increase is twenty-fivefold. A hundredfold collapse makes everything ten thousand times heavier. Things get out of hand at a rapidly increasing rate.

A normal star like our sun will shrink in its old age, but will stop when the star is about the size of Earth, at which point its individual electrons exert their braking pressure. The result is a white dwarf, a common but remarkably dense star. A baseball made of its material would outweigh two cement trucks. Its velocity of escape is several thousand miles per second, a frightening requirement that would discourage casual visitors or salespeople. Such solidity is awesome, but cotton candy compared to what happens when heavier stars start to shrink.

Stars weighing four, five, even ten times more than our own sun suffer from an unstoppable collapse. The forces set in motion are as irresistible as an avalanche. The smaller it gets, the smaller it wants to be, until the escape velocity reaches 186,282 miles a second, at which point light can no longer leave and the star effectively disappears from our universe.

In a way, nothing really changes at that instant. The star continues its catastrophic breakdown unaware that the outside world is now calling it a black hole. Indeed, the escape velocity continues past the magic light-

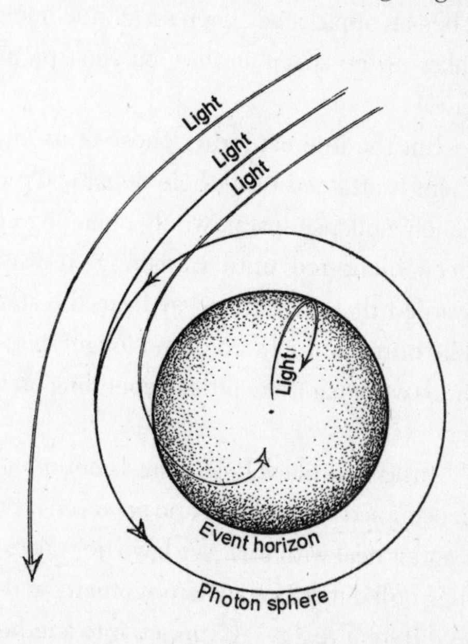

Light radiated within a black hole (center dot) cannot escape; it turns back at the event horizon. Light passing close to a black hole is forced into a curved path by gravity. If captured, it will either fall through the event horizon or orbit the hole at a distance 1½ times the radius of the event horizon (which is 1.86 miles for an Earth compacted to a quarter-inch radius). Objects slower than light will orbit farther away—in the accretion disk.

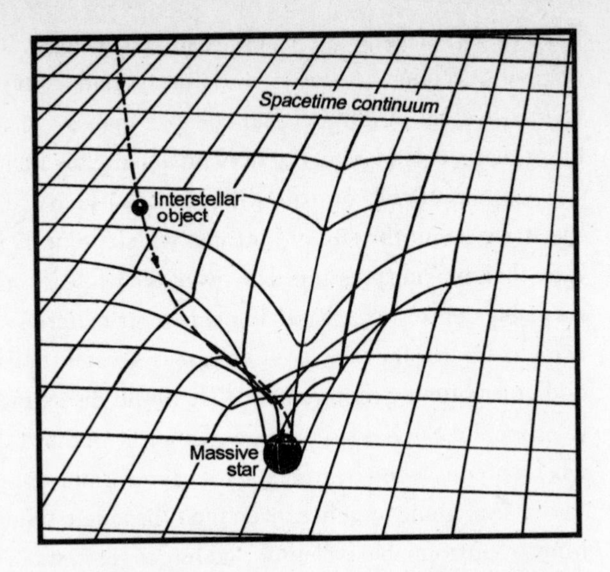

Gravity associated with a massive star warps spacetime much as a bowling ball distorts a rubber sheet.

speed number to 200,000, then 300,000, then a million miles a second. A billion. A trillion . . . as its density spirals upward to the point where a dust speck of its material outweighs a planet, and then a thousand planets.

And still it keeps shrinking. Cygnus X-1 achieved black hole density when it became 3.7 miles in diameter. Yet the star shrivels still further, until its mass of 3 million Earths is packed into the size of a beach ball. Then an apple seed. Then an atomic nucleus. And it keeps going, until it takes up no space at all. (You can't picture that? Neither can anybody else.)

But the interest for us, those of us on the outside, began at the moment it attained black hole density. Up until then it was merely an insanely collapsing star, wildly changing color from white to orange to a deep blood-red until vanishing. It was then that space became so warped that all paths away from the star folded back unto themselves like burritos. Light could no longer exit, and everything became sliced into two separate realities depending on whether you were on the inside or outside.

Inside the black hole, things continued to collapse in an eyeblink, until it occupied zero volume and achieved infinite density. Our laws of science cannot deal with this. We have no way to describe what things are like in this null dimension, where geometry and time become infinitely warped. A different reality? A "tunnel" into another era or age?

If the star was rotating—and they all do—then certain angles of approach permit hypothetical paths into other places or times. Enter exactly the right way and you're suddenly at the senior prom on the planet Zonda. Spacetime may pinch itself off from our universe, but it is impossible to know what may lie on the other side.

Nor should the fainthearted attempt the trip. Tidal forces would be so tremendous that if you fell toward the singularity (the zero-volume collapsed star at the center) your ankles would be yanked far faster than your knees, which in turn would speedily leave your hips behind. A ductile person, like the Rubberman of comic book fame, would find himself a miles-long strand of spaghetti. Less elastic individuals would instead be torn to pieces, each pulled forward at a different rate. But who wants to visit another dimension as a plate of goulash? Cancel the trip.

Outside the black hole, however, things would look different. First, the black hole would remain black even if you shone a powerful searchlight at it—although some light might go into orbit to form a ring, the **photon sphere.** Next, the strange way that time is distorted means that to an outsider, someone falling into the black hole would seem to slow down and freeze at the edge, hanging motionless forever.

Falling into a black hole is no problem at all . . . if you're made of rubber.

All this occurs at the event horizon, the invisible no-trespassing zone surrounding the collapsing star. When we think of entering a black hole, we really mean crossing the threshold of the event horizon, several miles outward from the crushed star itself. Step across it and you're doomed, for inside this region the escape velocity exceeds that of light. No matter how small the singularity at the center becomes, the event horizon remains a Venus flytrap. While nothing in particular happens to you as you cross the event horizon (except, as *Discover* magazine's Tim Folger pointed out, "Suddenly your phone calls aren't being returned"), your destiny is sealed. It's a one-way voyage from this point on.

If we could look at a black hole, this is approximately what we'd see. Near the center is the photon sphere, where light is trapped into an orbiting shell. Outside of that is the inner edge of the accretion disk, so hot it radiates completely in X rays. The accretion disk is actually flat, but the gravity of the hole bends the light from the far side of the disk over the top of the hole and upward underneath.

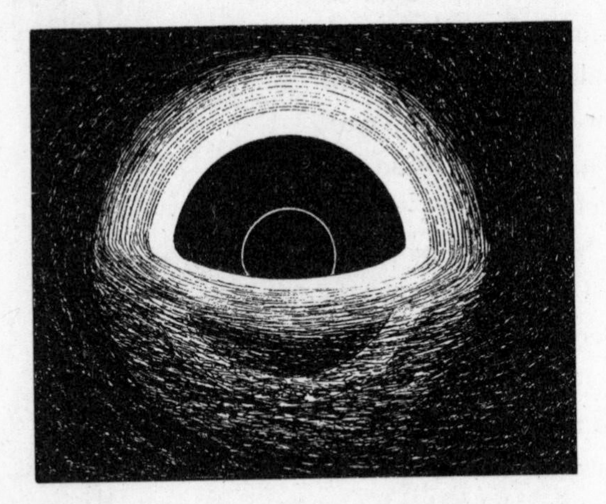

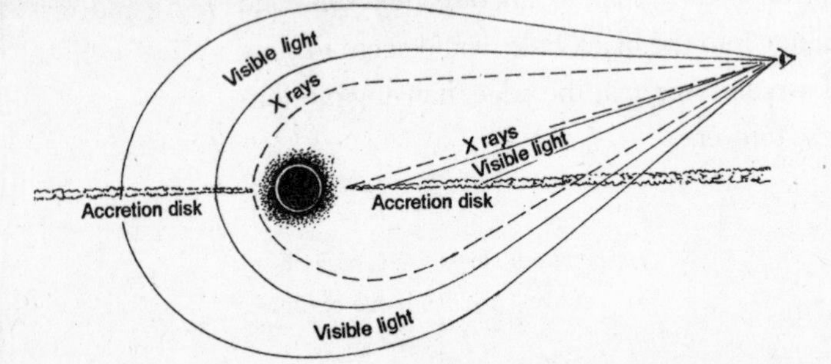

It all seems as intimidating as an IRS audit, until we recall that this entire black hole is just a few miles across and thus presents no great threat to interstellar navigation; it's like a tiny pothole on a multilane highway. Indeed, you could get as close as 2,500 miles—barely more than the width of the moon—to a ten-solar-mass black hole without suffering ill effects.

But such reassurances must battle perceptions, held for years, of black holes as shadowy parasites consuming the universe. In truth, if our sun collapsed to become a black hole (impossible, since it's not massive enough), we'd remain in orbit just as before, because its total gravity would be unchanged. Our lives would continue unperturbed, except for the lifestyle modifications we'd have to make due to the utter absence of sunlight and the end of all life on the planet. So, cancel the insurance: You're absolutely safe from being sucked into a black hole. To be pulled in, you'd have to go there practically begging for trouble, deliberately bringing your spacecraft to its dark doorway.

Surrounding the entire strange kingdom of a black hole is an accretion disk, a pancakelike zone where material from any companion star is accelerated to near-light velocities before spiraling into the event horizon. This is the terminal mailbox, the last place where anything, whether atom or astronaut, can commune with our own universe before permanently leaving it behind. When we detect the X rays alerting us to a black hole's existence, we are hearing the final frantic yelps coming to us from stardust caught in the accretion disk.

The core of our own galaxy, and perhaps every galaxy, may house a black hole that weighs not merely four to twenty times more than our sun, like typical stellar black holes, but a million or more solar masses. Black holes, small and large, may even account for some of the "dark matter" that astronomers now think composes most of the universe.

In fact, the universe itself might be a black hole! If it contains the "missing mass" many astronomers believe must eventually be found (and black holes of all sizes may well provide the unseen substance that seems to lurk everywhere, silently tugging at stars and galaxies), then our universe "positively curves" its own space, meaning that every path in each direction curls back into itself. If so, we live in a black hole: None of our light or material can escape. It illustrates how, far from being malevolent, black holes can be fun, what with amusement parks, fine restaurants, and everyone we've ever loved trapped along with us.

The down side is that residents of black holes are doomed to be crushed eventually into a sizeless singularity. Fortune cookies never mention this, but being squashed into subatomic oblivion is the ultimate destiny of every one of our descendants, and every planet out to the farthest galaxies—*if* our universe proves to be a black hole.

But that's not our concern today. We've got tens of billions of years before we reach our own crushdown, and by then, who knows? Maybe we'll have figured out how to navigate safely through the various-sized black holes of our galaxy, perhaps to arrive at other universes in different stages of their own evolution.

For tonight, Cygnus the Swan flies overhead, bejeweled with the mysterious black pearl called X-1. We can almost feel its bath of X rays teasing us, beckoning us toward a realm that lies not just outside our strangest dreams but beyond the stuff that dreams are made of.

Satellite Season

If you were born before 1950 or so, you can remember when only the moon orbited the Earth. Artificial objects circling the globe without further propulsion was a concept of science fiction.

To today's youth, that's not sci-fi but ancient history, and the October 4, 1957, flight of the first satellite is merely another textbook date, like Napoleon's retreat. The launch of that original Russian *Sputnik,* which sent the world to rooftops for an excited upward stare at the odd new roving point of light, is not the kind of thing that makes headlines anymore. Satellites are as commonplace as pizza. Few people regard them with awe and practically no one deliberately watches for them. Most folks say they've never even seen one.

How curious, then, the unfailing excitement that ripples through any outdoor group of novice astronomy students when a glowing dot glides noiselessly overhead. A good teacher can create astonishment by tossing off "That looks like a military spy satellite in low orbit, but it's no longer operational. Watch it vanish, before it even crosses the sky. There, gone!"

Now the audience is hooked. How could he figure all that out with just a quick upward glance?

It's easy. Satellite familiarity is as effortless as burning the toast. No unpolluted skies are required, no equipment, no ideal observing site. Not even much patience. However, there *is* a preferred time of year, a "satellite season."

It's now.

From May through August, dozens of bright outer-space machines cross the sky every hour.

More than six thousand satellites presently orbit Earth, and during the first hours of darkness each night—the ideal time to catch them—one will appear every minute or two. The ground rules are simple: If it's a slow-moving point of light and not a plane, it's a satellite.

But then what? Where do you go from there?

Here's a checklist.

Does it slowly zigzag? Most people are convinced, when viewing a slow-moving point of light, that it's weaving one way and then another as it crosses the sky. But satellites travel in laser-straight paths when seen from below; imperceptible eye-muscle movements cause this common illusion. In fact, when someone reports that a light is zigzagging we have a good clue toward identifying it as a satellite.

How fast is it moving? With a little practice you can easily recognize the different speeds that correspond to various heights aloft. We should remember that satellites are not propelling themselves. They are simply falling. That they can fall in a curve that matches the curvature of Earth below, allowing them to circle indefinitely, should not be too surprising. When you toss a stone or ball sideways, it too falls in a curving path—the result of gravity added to its horizontal motion. Toss it sideways fast enough (six miles a second) and the looping freefall would match Earth's curve. Do all this just above the atmosphere so that air friction doesn't slow the thing down, and you've got a satellite.

The satellite's speed depends on its height, to balance the pull of gravity. The higher it orbits, the weaker the gravity and the slower it will move. A typical military or communications satellite travels in a 250- to 600-mile-high orbit and will traverse the sky in a couple of minutes. If it's decidedly faster, it's a dead duck—a low, moribund satellite only 110 to 130 miles up, skimming through the drag of the upper atmosphere, soon to be pulled down to a meteoric destruction.

In short, a satellite usually crosses the sky about as quickly as a high-altitude commercial jet. This is an important clue for beginners. It immediately eliminates such common "UFO's" as meteors, which split the heavens in just a few seconds at most, and bright planets or stars, which scarcely move at all, even if passing clouds sometimes make them seem slowly in motion.

In what direction is it traveling? If it's coming from the east, that's persuasive evidence you're actually asleep and *dreaming* the whole thing. Because east-to-west is the one heading that's never seen. Satellites are launched the other way to match Earth's rotation. Planners take advantage of our planet's 1,040-mile-per-hour spin by sending the rocket eastward to give it an initial push, like a pitcher's dream of throwing his fastball from a speeding pickup truck. This confers more payload for less fuel: a free lunch. That's why all countries launch from as close to the equator as they can, where rotation is fastest. We didn't pick Cape Canaveral because of its proximity to Disney World, nor did the French choose Guiana because they like the wine there.

Some satellites occupy a polar orbit, instantly recognizable because they're heading more or less south-to-north or vice versa. This is a favored military trajectory because it allows the whole Earth to be scrutinized each day. Reconnaissance satellites tend to inhabit low orbits, so when you spy one with both high speed and a polar orbit, it may be spying you, too. Claims of satellites' being able to read the license plates of cars are exaggerated—but not by very much. Resolution of the U.S. KH-11 is probably in the half-foot range, allowing a passing satellite to see you taking a stroll, but not notice your bald spot or cellulite. Because of the inherent blurriness of the atmosphere, even sophisticated enhancement techniques may be unable to permit those final indignities until sometime in the next century.

Is its light steady or erratic? Remember, these sky invaders merely shine by reflected sunlight. In the old days of spherical satellites like *Echo* or *Telstar,* sunlight would bounce off them uniformly no matter what. Today's orbiting robots, with cryptic names like *Lacrosse,* possess complex shapes that reflect unevenly. If the machine is functional and stable, one side may hold the sun for quite some time. But in the case of a defunct satellite, the inevitable out-of-control tumbling causes obvious changes in brightness as it moves across the sky. As one watches, one can't help but visualize the dazzling glint of sun being thrown from the craft's shiny sides as it tumbles, causing the on-again-off-again effect of a toddler playing with a light switch.

Actually, such irregular special effects often come from satellites never intended to be functional; they're space debris such as third-stage rocket boosters that winked on and off from the day they reached orbit. Most objects tracked as "satellites" are just such space garbage, and their increasing presence has produced a girdle of scrap metal surrounding our world: the **junkosphere.** Luckily, most of this trash is small enough (less than a few feet across) not to catch our eyes in the first place.

But in some cases the rubbish is the whole show. When the world watched the first *Sputnik*, few were aware that they were *not* seeing the 20-inch-diameter satellite at all, which was generally invisible. The 100-foot-long body of the rocket's giant upper stage, which had also gone into orbit, was what everybody unknowingly stared at. The world was gazing with admiration at a giant unsalvageable piece of scrap metal.

A tremendous recurring headache has been the unfortunate tendency of the orbiting final stage of the U.S. Delta rocket suddenly to explode—sometimes after years of floating placidly in space. Unspent fuel is the culprit, leaking from gradually corroding pipes. One blew up in 1992 after being in orbit for fifteen years. The result: hundreds of pieces of debris, each large enough to require a catalog number *and to be tracked permanently as a separate satellite*!

Of the twenty-three thousand satellites tracked since 1957—of which a third are still in orbit—many are such debris, just a few inches in diameter. They are fortunately not our problem, as observers. Out of sight, they may as well be out of mind as well, for only some three hundred satellites are large enough and low enough to stand out easily to the naked eye.

When are they best seen? The visibility of satellites varies simply because of the changing sunlight that reaches them.

Obviously, the sun is shining in space even when it's nighttime here on the ground below. The moon, for example, always basks in sunlight; otherwise we wouldn't see it. No matter how dark the night, the sun's often shining "up there." But not always: When we look into the midnight sky we're staring right into our own planet's shadow, into a black sunless zone that extends into space for a million miles.

Today's satellites are irregularly shaped. Clockwise from upper left: Seasat, Gamma Ray Observer, Radio Astronomy Explorer, Infrared Astronomical Satellite, Landsat, Ultraviolet Light Observer, *and* Westar VI.

The nocturnal sunlight's presence or absence varies not only with the time of night but with the seasons. In winter the overhead sky quickly plunges into shadow after sunset. In summer, however, the shadow slants southward and never fully splays straight up, even at midnight. In warm weather, then, the heavens may appear fully dark by 10 P.M., but the sun quietly shines 100 miles overhead. The schoolbus-sized satellites catch that gleam and stand out against the inky backdrop like diamonds on black velvet.

What about that mysterious sudden disappearance? How could that teacher know the satellite would vanish before it even crossed the sky? Is such magic really predictable?

You can announce not only that the moving point will disappear, but even *where* in the sky it will happen. You have only to keep track of the invisible location of Earth's shadow in the night sky to know where any satellite must vanish. In summer, the shadow boundary starts out low in the southeast shortly after sunset and slowly rises to fill the mid-southern sky. Though the shadow boundary (the so-called twilight wedge—see page 227) becomes invisible to our eyes even before darkness has fully fallen, it's still there, and any satellite must disappear when it enters the zone where sunlight is absent at its orbital height. By 10 P.M. the shadow's edge sits 20 or 30 degrees (two or three fists at arm's length) above the

The MIR Space Lab *is still in orbit.*

southeastern horizon. In brief: opposite the sun. Its presence will be obvious, for the spacecraft will dim and vanish upon reaching it.

Finally we come to the most important satellites of all, as far as millions of people are concerned: the ones tracked by backyard TV dishes. Like pilgrims facing Mecca, an unblinking worldwide legion of antennas maintains a devotional focus on these same few dozen solar-powered transmitters. Utterly invisible despite their strong signals, they live high at 22,300 miles, some hundred times farther from us than the army of lower-orbit satellites seen crossing the sky every night. This renders them about ten thousand times less bright, putting them even beyond the range of binoculars. The last straw is that, being geosynchronous (revolving at the same rate as Earth's rotation), they're essentially stationary against the night sky, indistinguishable from a million background stars.

Beyond the two-hour postsunset period, happy hunting can also be expected a couple of hours before sunrise, a symmetrical interval of frequent satellite visibility ideal for insomniacs. (Remember, though, few will be seen from November through February, the season when our planet's shadow fully occupies the sky soon after nightfall. Satellite hunting is most productive just when the weather is most attractive, during the warmest third of the year.)

While identifying satellites and noticing whether they're polar, military, high orbit or low, operating or dysfunctional, you can also enjoy the idle pastime of seeing whether the moving point of light "touches" or converges with any star. Given the many faint stars that wallpaper the rural heavens, you'd think that the satellite would eclipse several as it traverses the bowl of night. The surprise: It rarely happens.

Only a few thousand stars are visible at any one time, and these are spread thinly among the 20,626 square degrees of the night sky. The fact that the moon would fit in a quarter-square-degree box speaks of the enormity of the celestial sphere. The satellites' surprising demonstration suddenly makes sense: that a straight line slashed across the heavens should rarely contact even a single naked-eye star!

Now that walking has been rediscovered as an exercise form, summer's plethora of satellites and its associated observational activities offer something to do while taking an evening stroll. With portable satellite-receiving equipment a good bet for the not-too-distant future, our relationship with these roving robots will probably become ever more intimate. That they're now so numerous makes them no less amazing. We may as well watch them—

They're watching us.

The Cosmic Background Explorer *behind the* Hubble Space Telescope

Struck by a Meteor

Only the extremely paranoid entertain a fear of meteors. The unlikelihood of being harmed has pushed celestial stones far down the list of most people's anxieties.

Perhaps such serenity isn't justified. In the early 1990's new calculations showed that a person is six times more likely to be killed by a meteor than in an airplane crash! This is true because when a really big one hits, possibly every hundred million years, it can destroy half the life on the planet in one shot. If you did all your traveling by plane—say by being extremely diligent about frequent-flier coupons—you probably wouldn't experience a domestic air travel accident for dozens of millennia. If that were the only thing that could harm you, you'd live practically forever. (Surviving so much airline food is another matter.)

If we confine our concerns to the present lifetime, the danger is reduced but not eliminated. After all, Mrs. E. Hewlett Hodges was not contemplating celestial thoughts on November 30, 1954, when a meteor crashed through her ceiling in Sylacauga, Alabama. Bouncing off a radio and striking her leg, it instantly granted her the dubious fame of being the only confirmed case in human history of meteoric injury.

(True, a Franciscan friar was reportedly killed in Milan in the seventeenth century by a two-inch meteor that severed an artery in his leg, but the confirmation isn't ironclad.)

Animals have fared worse. Newspaper accounts tell of a calf struck in Ohio in 1860 and a dog killed in Egypt in 1911. But the more realistic, ongoing threat is directed against property. It's been calculated that in North America alone, a house should be struck every 1¼ years, on average. Your home may be a fortress, but you probably didn't consider the risky piece of real estate you chose as its foundation. Nor did the Realtor fully disclose that your land hurtles through space at 18 miles per second, crashing annually through a dozen swarms of stone and metal.

Such considerations might now occur to Michelle Knapp of Peekskill, New York, who heard an explosive crunch from her driveway on October 9, 1992. Rushing outdoors, she found the rear of her car grotesquely smashed by a meteor, which lay beneath the crumbled trunk. While the required repairs lay decisively outside the manufacturer's warranty, such a mishap isn't necessarily bad news: A collector paid $69,000 for the ten-year-old Chevy—and the alien carwrecker.

That incoming meteor was witnessed by hundreds along the eastern United States, who saw the stone break into fiery fragments. This was not unusual; meteor landings, such as the dozens that impacted on a Ugandan village in 1992, are often preceded by widely observed celestial fireworks. My favorite meteor story involved just such a scenario in the northeastern United States on November 30, 1982.

An alarmed woman phoned my observatory that night to report a fiery ball slashing across the heavens, lighting up the countryside. People commonly assume observatories are UFO-reporting stations, and we get regular inquiries concerning lights in the sky. Like most, this had an easy explanation; I told the woman that the sparkling object was probably just a meteor, nothing unusual. We couldn't know, however, that things were anything but routine a mere 100 miles to our east.

Observers in central Connecticut were noticing the same brilliant light in the sky, but to them it was motionless. There was only one way it could appear stationary: It was coming straight toward them!

Surviving its fiery passage through the atmosphere, the grapefruit-sized meteor crashed through the roof of a house in Wethersfield, Connecticut, where Robert and Wanda Donohue were watching television—*M*A*S*H*—in the next room. Reacting to the cacophony, they found a hole in the ceiling, furniture knocked awry, and dust filling the air. For lack of an established procedure for this kind of thing they called the police, who brought along some firemen for good measure. One of them found the six-pound meteor under the dining room table, where it had settled after a couple of high-speed bounces chronicled by scuff marks on the carpet and ceiling.

Here's where the story gets interesting. Eleven years earlier, in April 1971, the last time a "shooting star" had impacted on a house anywhere in America, the community that gained that distinction was—Wethersfield, Connecticut. The same town! In the coincidence of the century, a house barely more than a mile from the Donohues' had been struck!

How could it happen? How could meteors strike the same town consecutively? And why Wethersfield? Did it have anything to do with its being a suburb of Hartford with its many insurance companies? Could a little cosmic prank have been played on the statisticians and actuaries who lived there and knew how high the odds against such a coincidence might be?

(And while we're asking rhetorical questions, was there any chance at all that the Donohues' insurance covered them for damage from meteors from outer space? The answer: yes, completely. And yes, again: The dual meteor bombardment of Wethersfield was strictly a bizarre coincidence.)

The publicity surrounding the curious occurrence, and the Donohues' generous donation of the celestial housewrecker to a museum in New Haven, did little to dispel the many commonly held meteoroidal misconceptions. For starters, they are *not* hot when they land. On August 31, 1991, two boys standing on a front lawn in Noblesville, Indiana, who heard a meteor thud into the grass a few feet away could touch it immediately. The icy lower atmosphere deep-freezes the stone so that it's only slightly warm by the time it reaches the ground.

Not that many do. Most visible shooting stars are the size of apple seeds, and of the 100 million that enter our atmosphere each day at speeds fifty times greater than a rifle bullet, nearly all disintegrate into dust. The bulk of meteoric debris thus falls upon our planet not as various-sized chunks, but as a fine powder that adds several million tons to Earth each year.

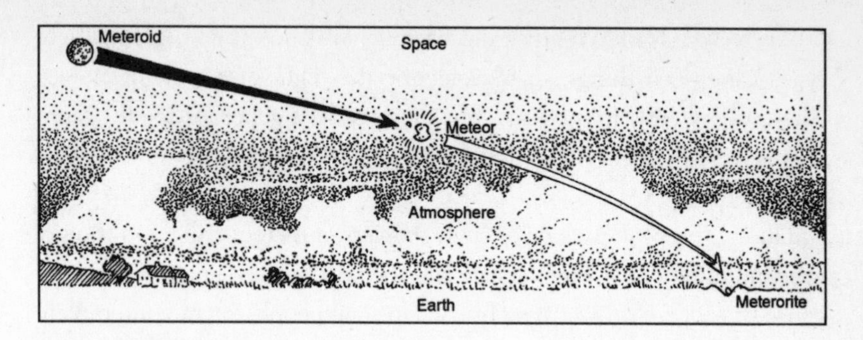

Quick name change

Another meteor quirk is a trait polygamists would envy—the ability to change names in a matter of seconds. Out in space the invisibly tiny, non-glowing chunks of stone and metal are called **meteoroids.** Upon hitting our atmosphere and being heated to incandescence by air friction, they're termed **meteors** (or shooting stars or falling stars). Upon making it to the ground, they become **meteorites.** No pesky grammarian can fault you, then, if you claim you were struck by either a meteor or a meteorite, since the conversion occurs at the moment of impact.

Meteors were given a mythological ability to grant wishes, but that's apparently insufficient incentive to get most people gazing at the sky. Many say they've never seen one. Nonetheless meteors are surprisingly common. One flashes across the heavens every ten minutes, on average.

Anybody may see a meteor any clear night of one's choosing. Impatient types can speed things up by selecting the night of a meteor shower. These vary greatly in intensity, but the two most generous are the summer Perseid meteors every August 11–12 and the frosty Geminid display on December 13.

During the peak periods of such showers, a meteor will cross the sky every minute or two. The Perseids are the best beloved and observed because they occur on warm nights, when sane people might contemplate remaining motionless outdoors.

Like debris tumbling from an overloaded garbage truck, most meteors begin their careers as fragments from passing comets. Every year when our planet intersects the comet's orbit (we return to the same region of space at the same date annually) we charge through the shower of confetti once more. Sometimes the original comet no longer exists; only its scattered particles remain as a luminous memorial to its passing. But at other times the comet's location provides clues to how rich a meteor shower is likely to be. In 1992, for example, the patriarch of the Perseid meteors, comet Swift-Tuttle, swaggered through our celestial neighborhood for the first time since 1862. The result was an enhanced region of rocky debris, and unusually rich summer meteor displays from 1991 through 1994—and probably beyond.

Things were even more impressive on November 17, 1966, when our world crashed through the disintegrated head of old comet Temple-Tuttle, which created a meteor *storm*. A hundred meteors *per second* (no, that's not a misprint) exploded over the skies of the southwestern United States, convincing many people that the end of the world was at hand.

The two richest meteor showers result from collisions between Earth and debris from comets.

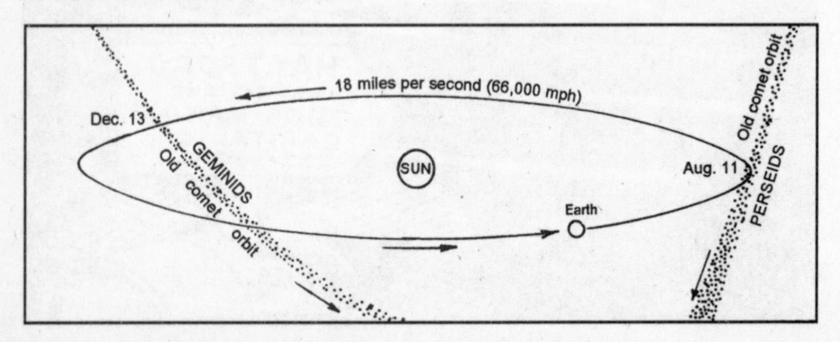

The meteor storm of 1966

One of them, a former student of mine, said she feared for her sanity as she watched the fireworks from the window of a train passing through the Texas countryside. As the celestial eruption intensified to the point where it seemed obvious that Armageddon had arrived, she looked around the compartment only to find that everyone else was asleep. She hesitated. What do you do? What's the correct protocol? Do you wake people up for the end of the world or let them sleep through it? Then the conductor came by and together they watched the unrelenting, bewildering exhibition until dawn erased the show.

Other firsthand accounts I've collected through the years echo the same impression of meteor storms: a spectacle beyond any other. Such a mind-numbing curiosity is expected to recur, over a narrow region of Earth that may be centered over the Middle East, on November 17, 1999.

The fact that most meteors derive from comets probably contributes to the public confusion about the two terms. Many use *meteor* and *comet*

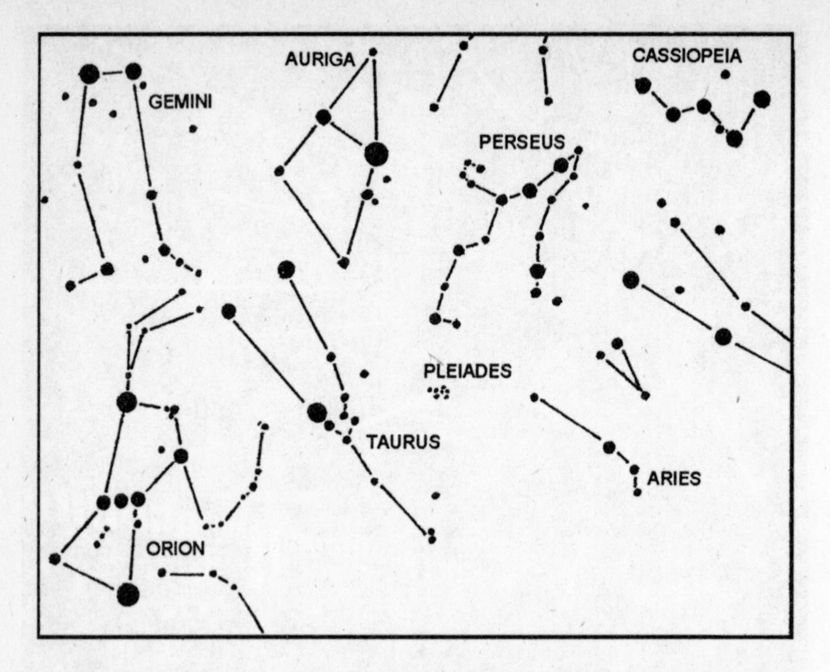

Where to look for meteor showers: Perseids emanate from the area between the W of Cassiopeia and the Pleiades, Geminids from the area above Orion's left shoulder.

as interchangeably as *groundhog* and *woodchuck*. But they're really very different.

Meteors are usually tiny grains passing less than 60 miles overhead, and vanish in seconds. Comets, on the other hand, lie as far from us as the planets, live for millennia, and stand nearly motionless in the night sky. Comets are followed by million-mile *tails* that persist for weeks or months, meteors by *trails* that rarely last longer than it takes to say "Look at that!" In short, if it moved and vanished it was a meteor.

Nor must their names be confusing. A comet adopts the name of its discoverer (one of the last celestial objects to keep that tradition; find a new star, planet, or asteroid, and an international commission will give it a name that will most assuredly not be yours). A meteor shower, however, is named for the constellation from which, by a trick of perspective, the meteors seem to radiate. If you plotted each meteor's path on the night of August 11, you'd see that most would be Perseids, meaning they all seem to radiate from the direction of the constellation Perseus. It's that simple.

Since Perseus rises around midnight, more shooting stars will be seen after that hour; you'll then view meteors radiating in all directions from Perseus instead of just the ones streaking sufficiently upward from that point to clear your horizon.

But talking about meteors is like discussing cuisine: The actual experience is much better. If you'd like to see the greatest possible display (and who doesn't enjoy the spectacle of a brilliant shooting star?) there are four rules for reaping the richest meteor harvest. They're simple and never change from year to year.

First, choose your location with care. Pick a spot with a wide expanse of sky. Don't just try to peek between trees and buildings. If only half the sky is in the clear, bring along a therapist to explain why you're deliberately cutting down the number of meteors by 50 percent. Lakesides, meadows, golf courses, rooftops, and deserts are all ideal. Even cemeteries, if you're so inclined. On a front lawn, turn off all house lights to allow your eyes the ten minutes necessary to become dark-adapted. If possible, get away from light pollution; this is a good time to visit those friends in the country.

Second, bring along a lounge chair to avoid neck strain, or spread a blanket. Take sweaters and mosquito repellent if necessary. Bring a friend and you have the recipe for an interesting and inexpensive date.

Next, leave behind your telescope and binoculars. The naked eye is best because you want to scan as much of the sky as possible. Just be sure the night is clear and not hazy or cloudy.

Finally, keep your eyes glued to the sky. Many people, out of habit, keep watching their companions during conversation and glance up only from time to time. This is a major meteor no-no. Looking at the person you're speaking to may be a standard rule of etiquette, but on that night the sky must be your paramour. This is particularly true during the August display. Perseids are among the fastest meteors because they collide head on with Earth—and seem to enjoy sneaking across the sky the moment you look away.

Meteors may be romantic, and bright moonlight has the same reputation, but here's a case where three is a crowd. Spend time with one or the other, but not both. Besides bad weather, the only natural factor that can diminish a meteor display is a brilliant moon. Ideally one wants a new

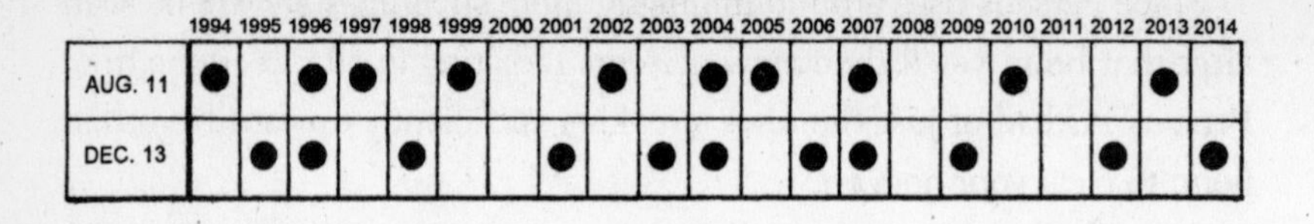

	1994	1995	1996	1997	1998	1999	2000	2001	2002	2003	2004	2005	2006	2007	2008	2009	2010	2011	2012	2013	2014
AUG. 11	●		●	●		●			●		●	●		●			●			●	
DEC. 13		●	●		●			●		●	●		●	●		●			●		●

Consumers' guide to meteor showers: The best years for watching are marked with bullets.

moon, the phase synonymous with darkness, but a crescent won't do much harm either. Even the first-quarter moon can be tolerated because it sets around midnight, which is when the Perseid show intensifies. A full moon is the worst news, not only because of its sky-ruining brilliance but because it's the single phase that is out all night long.

The Perseids start appearing each year at the end of July and build to a definitive climax on the night of August 11–12, after which the number of shooting stars rapidly declines. Perseids are most intense after midnight when your site on the spinning Earth starts pointing into the direction whence they come. They can therefore be spoiled by lunar phases from four days before full until three days beyond last quarter, when a bright moon is above the horizon during those critical hours. Good or excellent years for the Perseids will be 1996, 1997, 1999, 2002, 2004, 2005, 2007, 2010, and 2013.

The Geminid display has often been the year's richest since about 1920, but is underobserved due to the icy December 13 date. One meteor—or sometimes double that number—appears each minute *before* midnight when, conveniently enough, the display is most intense. The moon will keep its brilliance closeted, allowing good or excellent Geminid displays, in 1995, 1996, 1998, 2001, 2003, 2004, 2006, 2007, 2009, 2012, and 2014.

While meteorites are periodically recovered, especially in polar ice packs where they stick out like the alien intruders they are, you aren't likely to find survivors from either the Perseid or Geminid display. Cometary debris is more fragile than other meteor varieties made of iron or nickel, and almost always disintegrates into dust. You can watch it happen: About 30 percent of Perseids (but only 3 percent of Geminids) display lovely lingering trails as they delicately transmute into glowing skydust. Sometimes these luminescent trains magically persist, like the leftover smile of a special moment.

Watch for colors. These sky invaders, especially the sporadic meteors seen every clear night, commonly appear in orange, yellow, or emerald-green varieties. They may originally have been pieces of comets or asteroids, or even come from the moon or Mars. They are direct, dramatic reminders that we really do live in space: They are fellow creatures of the same forest.

And despite their potential for long-range disaster, the peril is minuscule on any particular night. So don't worry about being struck.

Unless, of course, you live in Wethersfield.

The Heart of the Galaxy

When the mood strikes, many of us like to go downtown for the buzz of activity and nightlife.

But where is the Main Street of outer space? What is the busiest part of the night?

It beckons us these summer nights. But paradoxically, to share in its crowded excitement and bright lights, urbanites must journey away from Earth's cities, whose skies offer a trifling floor show of just a few hundred stars. We must go far from human gatherings, to some isolated beach or meadow, where only the warm night breeze whirls between us and the riot of stars. Then the thousands of shimmering diamonds that adorn the rural heavens willingly introduce the dramatic band that dominates the summer firmament.

And there it is, more brilliant and complex, more mysterious than we remembered: the luminescent celestial highway many cultures considered the centerpiece of the entire sky—the Milky Way.

They were right. For as it cascades downward toward the south, it suddenly sports an extra-radiant section, like a splotch of spilled cream. This is nothing less than the center of the Galaxy!

The Milky Way, shown as it rises on early summer nights. The "Great Rift" of dark gas hides the glow of stars beyond. The Galaxy's center is at right.

More accurately, this glow marks the *direction* of the center, the stupendous, bulging heart of our spiral galaxy. We barely peer some 20 percent of the distance to the actual core before a profusion of stars and dust obscures the crowded turbulence beyond.

Radio and X-ray emissions have confirmed that the black heart of our galaxy is right there, beyond the constellation Sagittarius, whose stars resemble an "archer" only to those with gifted imaginations. Sagittarius actually looks like a teapot and is often labeled that way on today's charts. Adding the perfect touch, "steam" appears above the teapot's spout; it's this misty glow that marks the Galaxy's true center. But it's hardly a next-door neighbor. Those with an urge to visit that stormy galactic hub would have had to begin a light-speed journey when the pyramids were built, and continue beyond our lifetimes for another twenty thousand years.

What we *can* see makes it a showcase of the summer sky. Many astronomy aficionados wait until after the late news and talk shows are over, when the Milky Way's center is at its highest. Then they grab whatever instruments are available, wait a few minutes for the eyes to dark-adapt, and step outdoors. . . .

At that hour, the Milky Way bisects the summer sky from north to south like a bold artist's stroke. The galactic nucleus, fairly low and due south, seizes attention in inverse proportion to the sky's light pollution. In country darkness, particularly from southerly climes where it's high up, the intriguing glow of this "road to heaven" (as some ancients called it) pleads for exploration.

Try binoculars first. Except for the miniature "shirt pocket" models whose small lenses fail to do justice to the night, binoculars are superb for probing the Milky Way. Now every spot where one star shone before displays twenty newcomers in pastures dusted with intriguing nebulas and clusters. In dark skies, the binoculars' packed field of view overflows with countless stars scattered in a grand carelessness that defies description.

The ultimate may be experienced through the giant quarter-ton binoculars used by some Japanese zealots but rarely seen in the West. Fabulous views also materialize in ordinary telescopes if equipped with modern low-power, wide-field eyepieces. But strong magnification spoils the scene: What's desired is picture-window panoramas, not the few individual stars seen through the tiny porthole of high power. The public, whose general knowledge of optics is close to zero, often starts off with a megalomaniacal fascination with "power." Seasoned observers know how mistaken this is. They realize that to obtain a field of view crammed with

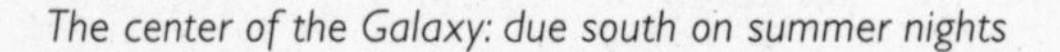

The center of the Galaxy: due south on summer nights

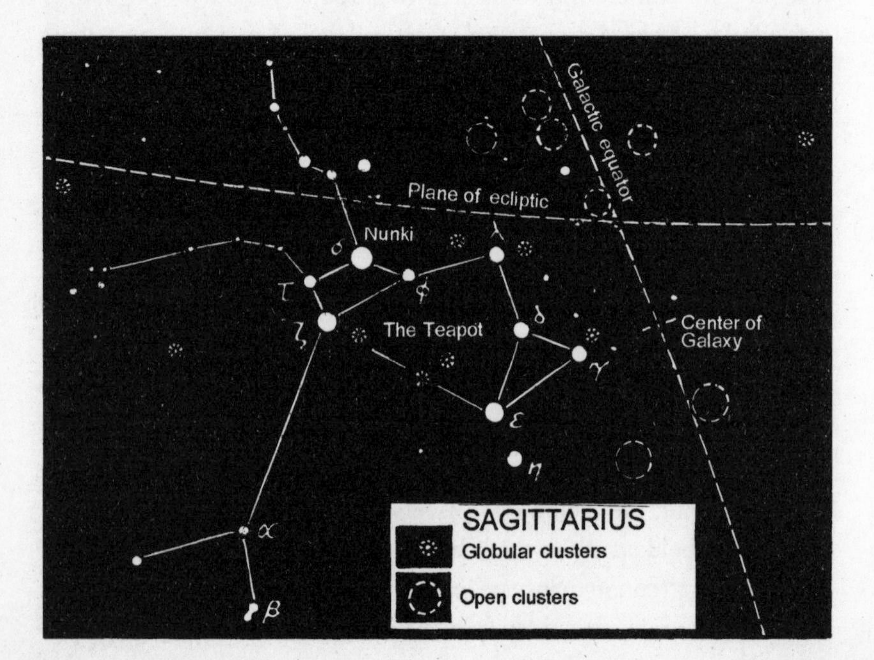

countless stars, the ideal instrument will employ a magnification of less than 60 ×. For example, 7 × 35 or especially 7 × 50 binoculars are excellent. Armed with such instruments or simply the naked eye, millions of amateurs around the world routinely experience the majestic ribbons of twisted dust and gas that float like ghosts near the Galaxy's heart.

What is it about the sight of multitudes of bright stars that so elates the human consciousness? Is it because our hugest home is space that our spirits feel intoxicated by the lunatic carelessness of endless stars in riotous, senseless formations? Whatever the reason, such "richest field" starry vistas are a major objective of summer stargazing.

Here, knowledge seems superfluous. Summer stargazers do not know the names of even one in a thousand of the stars that crowd the binoculars' field like an evening blizzard. Faced with such stellar extravagance, labels seem pointless. The nonverbal quarry is simply the curious, pancultural, uniquely human experience of wandering amidst endless anonymous points of light.

Such crowded splendor is not available in spring or late autumn, when our planet faces away from the Milky Way and toward intergalactic emptiness. Nor can binoculars provide such starry chaos when pointed in most directions in the sky. The experience is firmly denied those who observe from cities and large towns. Nor is it seen by vacationers whose trip to the country falls inopportunely during the week of the full moon, when the sky is robbed of its contrast and reduced to a handful of whitewashed stars.

No, for The View, no compromise is possible. The requirements are simple (even the naked eye is sufficient) but not amenable to negotiation. The sky must be clear, not slightly veiled by high, thin clouds. The air must be crisp, not hazy and humid. The site must be away from the lights of earthly civilization.

Here and there, jet-black streaks interrupt the glow. These dark nebulas, obvious to even the unaided eye, look as if some sinister celestial squid had sprayed splotches of ink on the Milky Way. Such colossal dust clouds add the third dimension; they lie somewhere behind the night's stars but in front of the Milky Way's fluorescent backdrop. Their cameo presence cloaks some of the Milky Way's glory, but also serves as an affirmation of continuous cosmic creation. These vast fields of interstellar dust—virtually absent at the core of the Galaxy but rashly abundant in the spiral arm in which we live—provide the raw materials for new generations of planets and suns.

So we're suddenly aware of a 3-D picture, where a moment ago the sky may have seemed flat, like the ancient notion of pinholes in the curving canopy of night. We catch on that the individual stars are nearest, even though they lie light-years away. Behind them are draped the immense blotches of dark matter that also make the Milky Way seem to split into two parallel streams like a divided highway: the Great Rift. And behind this dreamy darkness, peering out irregularly like sunlight poking through parting clouds, is the great glow itself.

The Via Galactica, as it was known in middle Europe, puzzled many civilizations. Typically they imagined it was the road walked by the dead on their way to heaven, or the messy mythological result of a feud between gods. Galileo was the first person known to point a telescope its way. Excitedly, he wrote that "wordy debates" about its nature could now come to an end, for in actuality it was composed of "innumerable coteries of stars."

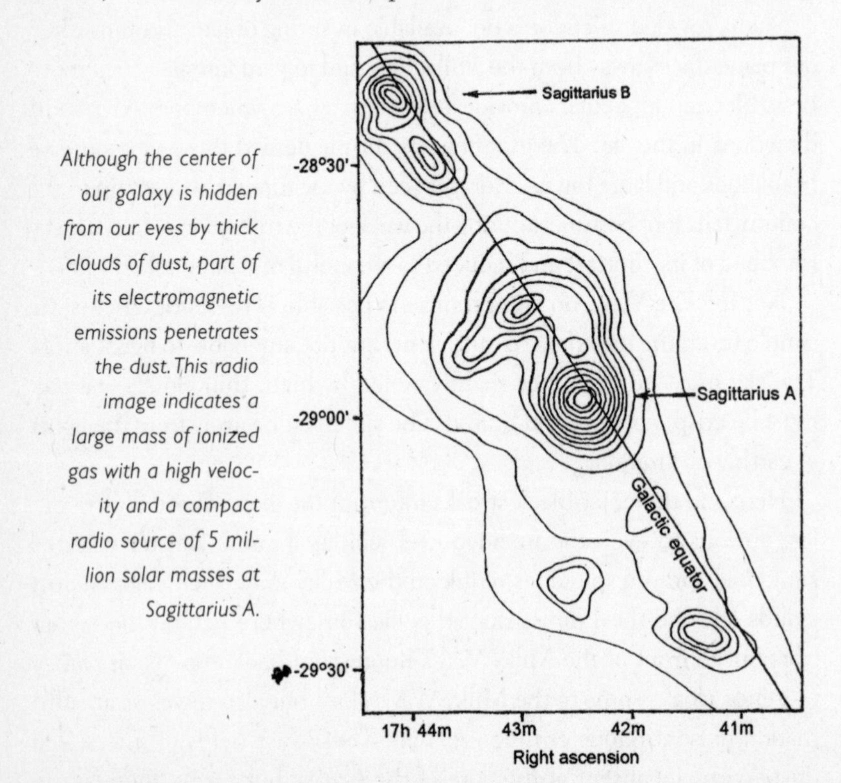

Although the center of our galaxy is hidden from our eyes by thick clouds of dust, part of its electromagnetic emissions penetrates the dust. This radio image indicates a large mass of ionized gas with a high velocity and a compact radio source of 5 million solar masses at Sagittarius A.

Indeed, so many billions of distant suns lie along the flat plane of our galaxy that their faraway radiance merges like melted gold to create this glowing band. The Milky Way encircles us completely; if Earth were glass, we'd see it as a ring fully surrounding us—our own disk-shaped galaxy viewed from our worm's-eye perspective.

Focus your binoculars and tens of thousands of stars spring into sudden visibility, the gentle radiance behind these newcomers indicating that still more unresolved stars lurk behind *them.* With every increase in telescope size, new stars materialize out of this light like dewdrops emerging from the fog, and yet the incandescence forever remains. More, still more, who knows how many billions more unresolved suns always lie farther in the distance.

Five thousand or six thousand light-years from Earth, the totality of stars and clouds of gas and dust becomes so thick it blocks the view of everything beyond. We see no farther. Only the penetrating light of wavelengths invisible to the eye breaks through, reaching our instruments to deliver hints about the crowded, violent heart of the Galaxy.

There is much indirect evidence that a large black hole occupies that central spot, whose mass may be at least one hundred thousand suns. Radio telescopes detect a double source of energy there, which has long been known as Sagittarius A and identified as coincident with the Galaxy's nucleus. A zooful of other odd celestial creatures lurk there as well. In 1993, Australian astronomers discovered a snakelike filament of gaseous material stretching for 150 light-years from the galactic center. Other smaller filaments, like Medusa's hair, curl outward as well—from the only place in the entire Galaxy where such strange structures are seen.

Our pancake-flat solar system of planets does not lie oriented in the same plane as the Galaxy, but is wildly tilted at two thirds of a right angle. While the Galaxy reveals its presence to our eyes as the encircling radiance of the Milky Way, the plane of the solar system also lights up—with planets, moon, and sun. The two planes intersect, by amazing coincidence, in the direction of the galactic core! Hence, foreground objects are routinely superimposed against the Galaxy's center. The moon visits this region of Sagittarius once each month. The sun faithfully marks the spot every mid-December. Jupiter brings our eyes to that direction with its dazzling light every twelve years, as in 1996 and 2008.

This means that even if your sky is too light-polluted to discern the Milky Way, you can face the center of the Galaxy merely by looking toward any June full moon, or by squinting toward the sun during the last two weeks of autumn.

If you have no desire to explore that region with optics, you might spend a moment simply staring. After all, half a trillion suns pay it homage. Every star of every constellation revolves around that spot of the sky each quarter-billion years, as the Milky Way performs its majestic rotation.

Including, of course, ourselves. The sun's grand circuit of the galactic center, bringing Earth unwittingly along for the ride, is sometimes called a **galactic year.** Perhaps a watch company with faith in its product will someday include a "galactic year hand" that completes one sweep every 240 million years, to tell us where we stand.

Until then, we'll stroll beneath the ancient summer glow, like generations before us, our self-importance diminished by the galactic vision that has only now come into focus.

Solstice of the Bull

The scene has become a cliché in adventure movies: A shaft of sunlight streams through a hole in the cave's ceiling, illuminating a clue to the treasure. But such scripts are inspired by fact. From Stonehenge to Mayan temples, the ancients loved to direct a beam of sunlight at a special nook on a particular day.

They often chose the solstice. All other days of the year have a twin, a second day when the sun's position duplicates itself. If your temple's idol was illuminated on March 2, the spot would light up again around October 9, so if you blew the ceremony you got a second shot. Those clever early astronomers realized that the solstice alone saw the sun occupying a parcel of sky unshared with any other occasion.

If your house has a window with an unobstructed eastern or western horizon, you know how fickle are the rising and setting points of the sun. In spring or fall the sun jumps nearly a solar diameter to the left or right of its position the day before. This motion slows about a month before the solstice, and comes to a dead standstill on the solstice itself. At that point, at its extreme, sunlight rushes into the room at an angle not seen at any other time.

As the media love to remind us, the June solstice marks the start of summer and the longest hours of sunlight. Skygazers take it a step fur-

Watching the sunrise on the winter solstice from within a centuries-old structure oriented for that purpose can be a moving experience. This is one of several such structures in Putnam County, New York.

ther by observing the year's highest sun at 1 P.M. that day, and by watching it set as far to the right, or north, as possible. The old myth about a north window never getting sun then gets exploded: Watch it stream in obliquely during the first and final hours of the day. (A north-facing window will see no sun whatsoever only from September to March.)

But why should the ancients have all the fun? Why not set up your own solstice device? Just notice where the extreme edge of sunlight strikes the room early or late on the day of the solstice. That exact spot won't get direct light again until you're a year older. It's a site just begging for its own idol.

Aging hippies might hang a small prism there, to let a rainbow of color mark the occasion. Gadget lovers could install a solar cell to power a chime or bell. A photoelectric cell could turn on a stereo, preset with the Beatles' "Here Comes the Sun." Those Mayans could build fabulous pyramids, but we could teach them how to lighten up.

Viewing our planet from space, you'd see that the solstice is the moment when the top or north part of Earth's tilted axis is pointing most toward the sun. Astronomers can predict this occurrence to one-second precision. The following table lists the instants of both summer and winter solstices for the years surrounding the turn of the century.

Solstices

Times are given in Greenwich mean time; 13 hr. = 1 P.M. Deduct four hours for eastern daylight time, seven hours for Pacific daylight time.

	JUNE	TIME			DEC.	TIME		
1995	21	20 hr.	35 min.	26 sec.	22	08 hr.	18 min.	15 sec.
1996	21	02	24	52	21	14	07	21
1997	21	08	21	06	21	20	08	34
1998	21	14	03	45	22	01	58	01
1999	21	19	50	20	22	07	45	21
2000	21	01	48	55	21	13	39	0
2001	21	07	38	56	21	19	23	04
2002	21	13	25	36	22	01	15	57
2003	21	19	11	38	22	07	05	21
2004	21	0	58	04	21	12	43	11
2005	21	06	47	20	21	18	36	31
2006	21	12	27	03	22	0	23	42
2007	21	18	07	39	22	06	09	25

At the summer solstice, the South Pole's pole (yes, they've really erected a pole there!) stands shrouded in its deepest darkness, while the North Pole basks in the year's loftiest sun. Here, you'd witness the sun's still anemic height as equal to our planet's tilt: 23½ degrees. You'd also probably marvel at its perfectly sideways motion around the whole sky,

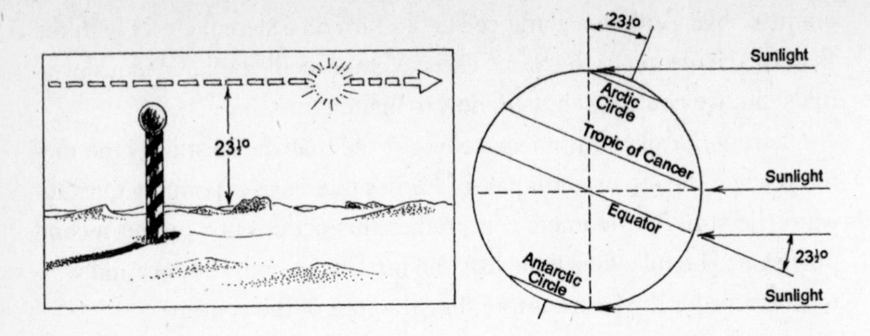

Left: *Summer solstice at the North Pole. The sun appears to move horizontally.* Right: *The Arctic and Antarctic circles, like the tropics, are defined by the sun's reach at the solstices.*

displaying not the least interest in rising or setting. Standing as high as an outstretched hand from thumb to pinky, and becoming neither higher nor lower, the sun chugs rightward hour after hour, day after day.

The late astronomer George Abell told me of a botanist who thought the North Pole a good place to raise sunflowers, since the sun shines there for months at a time. As recounted in his excellent textbook *Exploration of the Universe,* the flowers did very well for a while, but, "since they like to face the sun, and as they followed the sun around and around the sky, they ended by wringing their own necks!"

Quietly and with little hoopla, the solstice has just started occurring in the constellation of Taurus (meaning that if the background stars were

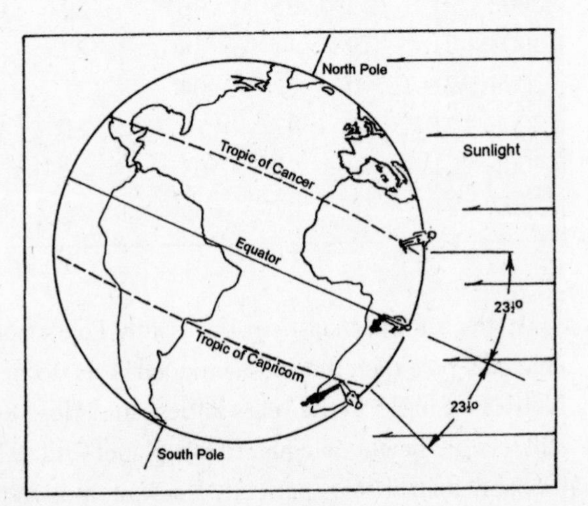

On June 21 the sun is directly overhead for people on the Tropic of Cancer.

visible by day, Taurus would be seen behind the sun on the date of the solstice). Until the 1990's, and going back to the time of Christ, the solstice happened in Gemini. For the next three thousand years the solstitial sun will remain firmly entrenched in its new bullish home, thanks to our planet's leisurely 25,800-year wobble. But *astrologers,* who ignore precession and the actual constellations, place the solstice in the "sign" of Cancer, just as in Ptolemy's day. Such living in the past explains why the region of Earth where the solstitial sun shines straight overhead continues to be called the Tropic of Cancer. By rights, the people down there should change the signposts to read TROPIC OF TAURUS. (But perhaps they're too stubborn.)

Tilts of revolution and rotation

The inclination of each planet's orbit relative to Earth's is shown to the left of the planet. The tilt of the planet's equator is placed by the pole that rotates as our North Pole does (arrows).

Since everyone in the contiguous United States lives north of the Tropic of Cancer, no mainland American city ever gets to experience an overhead sun. But in southern states it's a near miss at the solstice. How close does *your* shadow come to disappearing? You can quickly tell by how many degrees the sun misses the zenith in your location by subtracting 23½ from your latitude. Key West, at latitude 24½, thus sees the midday sun only 1 degree away from straight up, while Chicago's 42-degree latitude never permits the sun to venture closer than 18½ degrees from the zenith.All this—solstices, seasons, change in length of daylight—happens solely because our planet flies slantedly through space, the way Charlie Chaplin walked. If we visited Jupiter or Mercury, which stand almost perfectly straight up as they orbit the sun, we'd find no seasons at all and no change in weather throughout the year, except for the temperature variations caused by fluctuations in distance to the sun.

But we're not the oddballs; *they* are! Half the planets have tilts that are 20-something degrees, so that our sort of seasonal variation is the norm. The really wild situation occurs on a world we don't spend much time thinking about—Uranus. Its 98-degree tilt gives it amazing seasons.

During Uranus' leisurely eighty-four-year orbit, a summer day lasts for an idyllic forty-two years. During that time, you'd find the sun always out, the flowers always in bloom. (Okay, I'm exaggerating about flowers. There's not even a surface on Uranus; it's a thick green gaseous ball.) Then night falls for another forty-two years, the longest night in the known universe—the stuff of cabin fever.

On some worlds, the eccentricity of the orbit combines with the seasonal tilt to yield striking consequences. Even here on Earth, which neither has an orbit that's wildly out-of-round nor displays an extreme tilt, the effect is important, if relatively little known. The difference between our closest approach to the sun, in the first week of January, and the farthest point, just two weeks after the summer solstice, is about 3 million miles, no small potatoes. Since the intensity of light changes with the square of distance, the sun appears 7 percent dimmer in July than in January.

Imagine if things were different. What if the sun were 7 percent weaker just when our own Northern Hemisphere was pointing away from it, in winter? Wouldn't that make winters much colder than they are?

It would. And that's just what things were like thirteen thousand years ago, and will be like again in the future, when our planet's tilt faces the other way. When you think about it, that's the story right now for people in the Southern Hemisphere. Australians find that now, during their winter, is when the sun is most distant, making their climate significantly colder than it would be otherwise.

Fortunately for them, the Southern Hemisphere has much more oceanic area than the Northern; the tendency of water to heat and cool more slowly than land moderates the climate enough to make summers and winters pretty much comparable in both hemispheres. But stick around; in the distant future when our part of the globe points into the sun at the same time that we're closest to it, you'll *really* need air conditioning in the car. Winters will be extreme as well. Figure in the fuel taxes that the government will have imposed by then, and you'd yearn for these good old days thirteen thousand years in the past, when the worst that would happen was war, famine, and interminable reruns of *Gilligan's Island.*

The summer solstice beckons us to travel. Tourists trek to places like Lapland to see the **midnight sun.** It's an easy phenomenon to understand when we picture the North Pole with the sun circling the zenith and neither rising nor setting. Travel away from the pole, as you'd be inclined to do sooner or later, and the spot the sun appears to circle becomes increasingly shifted into the northern sky. The solstitial sun remains circling the sky all twenty-four hours of the day as long as you don't venture more than about 1,700 miles from the pole. When you reach that place—the Arctic Circle—the sun's daily circle is offset enough to hit the northern horizon at its low point. Travel farther south and increasingly more of its path lies below the horizon.

Figuring in the refraction of the sun's image, the midnight sun can be seen from as far south as latitude 66, in mid-Alaska—a delightful experience were it not for the countless galaxy-class mosquitoes thriving on the summer water that cannot soak into the ground because of the permafrost a few feet down.

Come to think of it, we can probably appreciate the solstice as well from home, and just as much as the ancients did—even if the law does discourage the exuberance of an occasional human sacrifice.

Eclipsing It All

Three glorious phenomena result from a peculiar feature of the sky itself, a coincidence so stupefying that it has been used as a credible argument for a Divine Plan. Yet, relatively few people are aware of this obvious but bizarre property of the heavens.

To the naked eye, the sky is simple. It's basically an inverted bowl hosting thousands of dots and two luminous disks. The points—stars and planets—exhibit no size because of immense distance. The two disks are sun and moon.

The question: Having seen sun and moon since early childhood, can *you* recall which appears larger?

It's surprising that 98 percent of a population that can differentiate the facial characteristics of hundreds of friends, acquaintances, movie stars, and the like cannot remember the dimensions of objects seen repeatedly for years. If our sky appeared the way it does from moon-rich Jupiter, Saturn, or Uranus, and displayed more than a dozen disks, it would be understandable to lose track of their relative dimensions. But two?

The unlikely truth: The sun and moon appear the same size.

The sun is four hundred times larger, but also four hundred times farther from us. This fact alone allows the moon to fit perfectly over the sun's

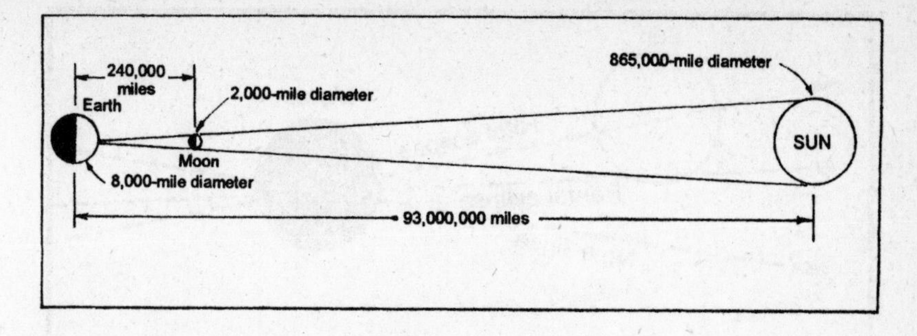

By sheer coincidence, lines drawn from the top and bottom of the sun through the top and bottom of the moon meet—at Earth!

face during an eclipse. Not too much to block out the beautiful inner corona, and not too little to leave the sun's blinding photosphere uncovered. This bizarre coincidence does not hold for any other planet, and indeed was not so here on Earth prior to the advent of humans, thanks to the moon's gradually increasing distance as it spirals away like a bent skyrocket at the rate of an inch a year. And if we are to explore solar eclipses, those most awesome and unlikely works of nature, we may as well do so in this season when the sun is most prominent.

That the two disks in our sky appear to sport the same dimensions produces several instant spin-offs, visible only when they align perfectly.

It doesn't happen often because of the moon's 5-degree orbital tilt; it misses the sun almost every time it passes by. And even when it does block the sun, the moon's shadow, tapering like a chopstick to barely

The moon's orbit is inclined 5½ degrees to the plane of Earth's orbit, crossing it at two points called nodes. Unless a node comes between the sun and Earth, an eclipse cannot occur—which is why the shadows of Earth and moon miss each other during most new and full moons.

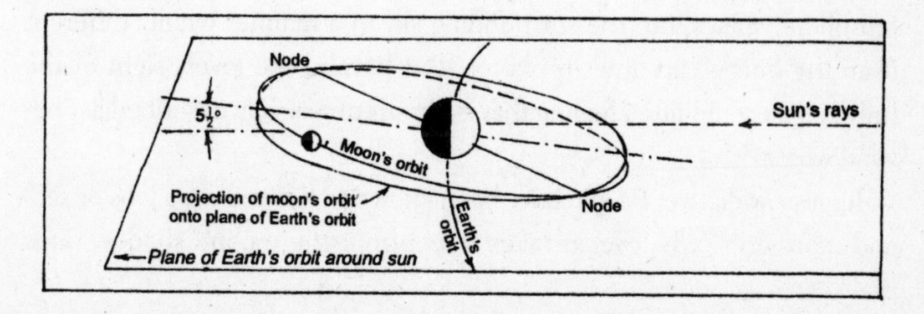

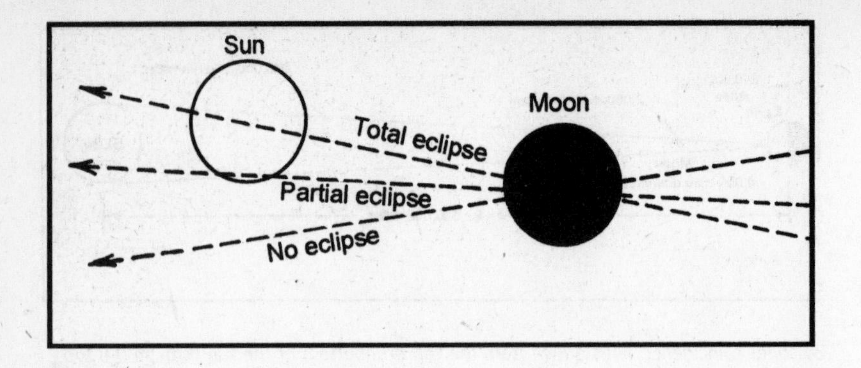

In a total eclipse, the center of the moon passes over the center of the sun.

touch the ground, can be fully viewed from just a tiny section of Earth. If you're lucky enough to be in the right place at the right time, you experience a total solar eclipse, one of the greatest spectacles the eye can behold. From a much wider area it's merely a *partial* eclipse, a fairly common sight that requires eye protection.

You'd think a partial eclipse would be almost as impressive as totality. But it's not even close. The difference is day and night.

A unique visceral *presence* apparently develops when the sun, moon, and your spot on Earth form a perfectly straight line in space. Animals aren't alone in exhibiting aberrant behavior; a total eclipse makes many people shout and babble as if the event were an excursion by asylum inmates. It's estimated that about half of all observers make such primal, unsophisticated sounds.

Moreover, the experience ineffably surpasses the merely visual. Certainly it is very different than it appears in photographs, which, because of under- or overexposure, cannot capture the same vast range of brightness as the eye. The delicate otherworldly tendrils of the sun's corona, or atmosphere, splays into the surrounding sky in a manner wholly different from the image captured in photos. But beyond the lovely sight of the fully eclipsed sun lies a feeling that is almost universally described as "beyond words."

It's also addictive. People have been known to assume extra jobs or second mortgages, whatever it takes to confront the moon's shadow once again.

The moment of truth for eclipse chasers can barely be hinted at in an artist's rendering. The fuzzy edge of the moon's shadow moves through the atmosphere at left.

Totality, lasting between one second and seven minutes, may be the world's most expensive indulgence. If the average observer has paid perhaps $1,800 to travel to a three-minute event, totality costs about $10 per second! Fortunately there are several bonuses. Two rank as among nature's most amazing phenomena all by themselves.

The first is the peculiar appearance of the surrounding countryside during the ten minutes prior to totality, when the sun is more than 80 percent eclipsed. The commonplace then becomes extraordinary; ordinary streetlights and buildings exude an exceptionally strange presence. Since sunlight now arrives only from the edge or limb of the sun—a condition normally unavailable—surroundings take on an unearthly quality, as if a different type of star were providing the illumination. Colors turn richer and more saturated. Shadows mutate, becoming oddly crisp and sharp. Thousands of glowing crescents, like Cheshire Cat smiles, appear in the shadows of trees. Like the distant laughter of a woman in a dream, an overwhelming sense of something truly alien establishes itself.

Glowing crescents in the shadow of a tree

By this time the steady drop in temperature usually produces an **eclipse wind.** Its gentle howl adds another haunting touch. But the best is yet to come.

A minute or two before totality, all light surfaces, such as the sidewalk or beach you're standing upon, suddenly exhibit **shadow bands.** Dramatic, shimmering dark lines that look a bit like the wiggles at the bottom of a swimming pool are now everywhere. This eerie phenomenon, which can make the hair on one's neck stand on end, *cannot be photographed.* Try it! Your pictures will show the scene without them.

Only recently explained, shadow bands are now believed to be the edges of atmospheric temperature cells (air pockets) uniquely projected by the tiny remaining point of sun, a bit like the shimmering lines of distorted air that appear over a radiator. Their dramatic motion readily catches the notice of the eye despite their extremely low contrast.

The brighter stars come out during totality and—another surprise: They're reversed! If it's summer, the winter constellations emerge.

Where must you go to witness such extraordinary apparitions in the same half hour—the oddly illuminated countryside, the unphotographable shadow bands, totality, and the rest?

You must make a pilgrimage. A total eclipse happens only once every 360 years, on average, for any given part of Earth. And if it's cloudy, you wait *another* 360 years. True, if you live in southern England, you can stay put and catch the one in 1999. But most of us must undergo a distant trek in order to stand in the moon's shadow.

Unfortunately, eclipses seem to display a perverse attraction to such hard-to-reach places as the Antarctic or Siberia. They're also famous for nerve-wracking last-minute clouds, ever threatening to obscure the whole thing. Bring more than just No. 14 welder's goggles for the partial phases. Bring antacid tablets, too.

If it's cloudy, the observer is reduced to staring idiotically into darkness, which is what the uninformed imagine a total eclipse to be like in the first place. Of course, the attraction isn't simple blackness. If it were, astronomers could stay home and get the same experience by disregarding their electricity bills.

The adventures befalling eclipse chasers could fill volumes. There are tales of sudden ruinous clouds, and happier stories of observers whose stalled vehicles deposited them by chance at the one spot where the overcast parted like the Red Sea to uncover the mind-numbing spectacle. Here's a typical example, from among hundreds I've heard:

In 1981, a group of Americans en route to an island in Lake Baikal, where the "ideal" eclipse site had been prepared by their Russian hosts, found themselves missing two members of their party. This pair of astronomers, by oversleeping, had missed the ferry and had to resign themselves to watching the event from the hotel roof.

But as the eclipse approached, a single stationary cloud formed over the lake. The only Americans to view the event were the late sleepers. What kind of moral can be drawn?

The following table lists every upcoming total eclipse until the end of the next decade, along with lengths of totality and a guess about the prospects for clear weather at the best location. Obviously, if you can afford to travel only once per decade, you don't want to go to one that occurs during a region's monsoon season. Conversely, I agreed to be "eclipse astronomer" for an expedition to Baja, Mexico, in 1991 because it was the clearest place on the continent during that month, according to long-term records. Many others learned about weather statistics the hard way by going to Hawaii, which was largely overcast.

Total Solar Eclipses

DATE	LOCATION	WEATHER	TIME (MIN.: SEC.)
Nov. 3, 1994	S. Peru, N. Chile	fair	4:23
Oct. 24, 1995	N. India, Thailand, SE Asia	good	2:10
Mar. 9, 1997	Siberia	fair	2:50
Feb. 26, 1998	Colombia, Venezuela, St. Kitts	good	4:09
Aug. 11, 1999	Western to SE Europe	good	2:23
July 21, 2001	S. Africa	fair	4:57
Dec. 4, 2002	S. Africa, Indian Ocean	fair	2:04
Nov. 23, 2003	Antarctica	poor	1:57
Apr. 8, 2005	S. Pacific Ocean	fair	0:42
Mar. 29, 2006	Africa, Turkey	good	4:07
Aug. 1, 2008	central Russia, Mongolia	fair	2:27
July 22, 2009	China, S. Pacific Ocean	good	6:39
July 11, 2010	S. Atlantic Ocean	fair	5:20
Aug. 21, 2017	next eclipse in U.S.	good	2:40

Notice that the first decade of the new millennium is not exactly an ecliptic Shangri-La; many occur in semi-accessible areas. The ensuing decade is even worse, with most eclipses falling over polar or oceanic regions. A notable exception is the coast-to-coast U.S. eclipse of 2017.

Such long eclipse droughts are common, but not invariable. In the four-year period beginning in April 2024, for example, we'll have three long, accessible totalities (4½, 5, and 6½ minutes) occurring over the United States, Spain, and Australia. It underlines the importance of each attainable eclipse, such as the duo that finish out the twentieth century.

Aside from the vicissitudes of weather, labor rather than luck determines who will be knighted by the Wonders of the Grand Coincidence. For its enchanting apparitions are granted only to those who, following nature's clockwork lead, take the time and effort to rendezvous with the shadow of the moon.

Twilight Time

Heavenly shades of night are falling,
it's twilight time.

—1958 hit song, "Twilight Time"

That tune by the Platters would not have made the charts on Pluto. In fact, creatures on any known planet would be amazed by the unique twilight we earthlings take for granted. Our world alone sees the riotous palette of colors that marks day's transition to darkness. We alone witness its wonders, whose very names—green flash, twilight wedge, zodiacal light—invoke otherworldly visions.

Throughout most of the universe, the sun sets and . . . wham, it's a power failure, a bewildering snap into blackness. Even when the transition from light to night takes longer, like on a slowly rotating or cloud-covered planet such as Venus, it still generates no color. The moon's experience is typical: Day or night, its sky never deviates from starry darkness. The sun always appears surrounded by black sky and stars. Sunset unfolds for a leisurely hour, but the heavens remain unchanged. Of the universe's known planets, only Mars offers an anemic attempt at twilight tints. Its thin air, however, is incapable of reproducing Earth's rich hues. That leaves just us earthlings to ponder the phenomenon, whose witnesses include bats and other crepuscular characters who cleverly avoid both day and night predators.

(*Crepuscular*? It means "pertaining to twilight," a wonderful word that's fun to use as often as possible, even when only marginally appropriate.)

In poetry and literature, twilight has a sort of transient, otherworldly status, as if a mere transition zone: "The dreamcrossed twilight between birth and dying," said T. S. Eliot. Most people see it that way, as an ephemeral buffer between day and night rather than as a lavish self-contained province of its own. This dreamy reputation probably stems not just from its brevity in many places, but from its vagueness. Twilight suggests something shadowy and indefinite.

In reality, twilight is a very specific event. Laws, and occasionally lawsuits, have been structured around its precise arrival and departure. And there's not one twilight but three!

Civil twilight starts at sunset and ends when the sun has plunged 6 degrees below the horizon—a dozen times its own apparent diameter in our sky. That's when streetlights must be on, according to most municipal ordinances. Enough light remains to stimulate the eye's color receptors, and you could still easily read this page.

Nautical twilight persists longer, until the sun is twice as far down, or 12 degrees below the horizon. That's when the horizon vanishes,

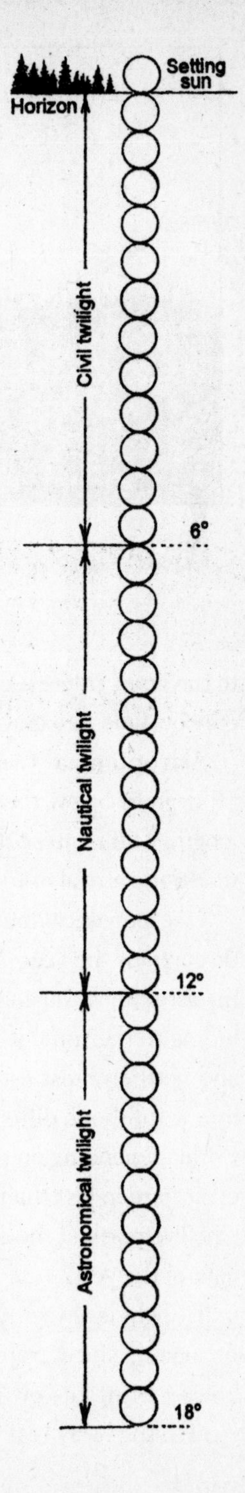

The distance the sun must plunge below the horizon to produce the three different types of twilight

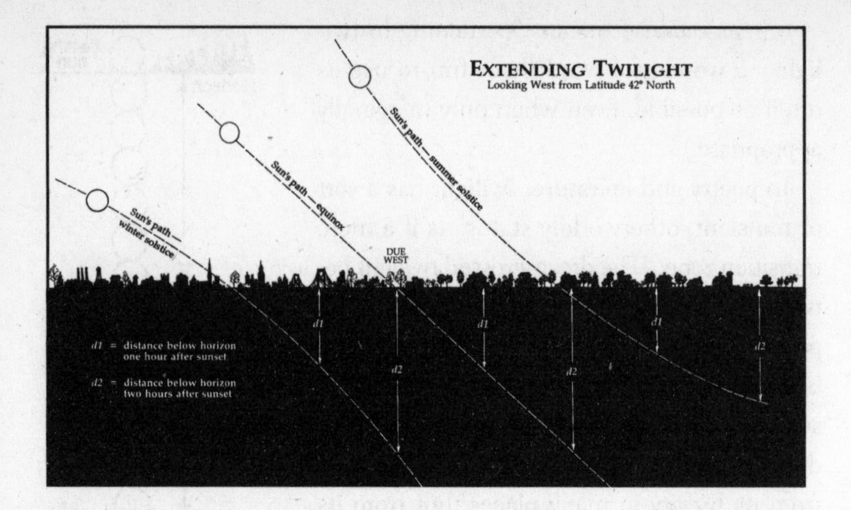

Twilight is longest in summer because after sunset the sun's path is more horizontal than vertical. The sun creeps to the right just below the horizon, keeping the sky bright.

to the point where a mariner can no longer distinguish sea from sky. Colors now fade into grayness.

Astronomical twilight continues still longer, until the sun has fallen 18 degrees below the horizon, or about two clenched fists held at arm's length. The faintest stars now emerge. The end of astronomical twilight marks the arrival of full darkness, the official commencement of night.

Like a mythical mutating creature, twilight's onset and duration change throughout the year. Moreover it's a different story for each latitude. In my astronomy musings on Northeast Public Radio, I take care when alluding to the time of nightfall; the variation between our southernmost and northernmost listeners, who live merely 200 miles apart, translates into a full-hour difference in the length of twilight! Elsewhere in the world, depending on the season and the location of the observer, twilight can expire in less than an hour or linger throughout the night. For people who live north of the equator, May through July brings the lengthiest twilight of the year.

Twilight is always shortest in the tropics, where one hour is all you get. A shame, since tropical twilight with swaying palms and ambrosial breezes brings its own special magic. But such romantic fantasies must yield to the hasty reality of rapid nightfall. Night comes quickly there because the equatorial sun sets vertically, swiftly dropping like a stone below the horizon. In our temperate regions the sun glides into the horizon at an angle, slipping sideways as it descends. An hour after sunset the sun's not very far down because it's moved rightward as much as downward. This has been obvious since antiquity, for anyone can keep track of the vanished sun's movement by watching the brightest area of twilight slither rightward as it broadcasts the sun's subterranean position.

From the latitude of Chicago, the average duration of daily twilight is about three hours. A mere 500 miles farther north there simply is no night at all in early summer: Evening twilight merges into the morning version to produce a permanent summer skyglow.

This total absence of summer night occurs in all places north of about latitude 50 degrees, which includes every country from England and Poland to Russia and Norway. In such lands twilight is a more persistent companion than night itself, through much of the year. So much for its being a mere transition period!

Twilight is distinguished by colors that are simply displays of various sections of the sun's spectrum. No mystery or alchemy is involved. The light from the low sun, arriving at your site at an angle, must penetrate more air; when the sun is 30 degrees above the horizon, sunlight passes through twice as much atmosphere as when it's overhead. This gives the shorter (blue) wavelengths more opportunity to exercise their tendency to

This side view of Earth shows why twilight lasts all night in the summer above 50 degrees north latitude. At midnight on the solstice, the sun is only 17 degrees below the horizon.

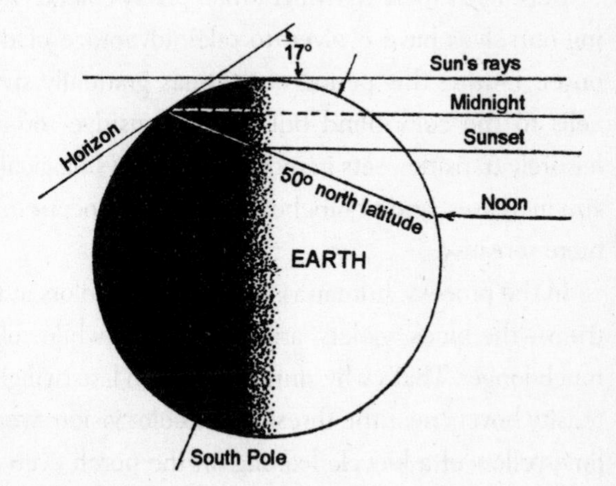

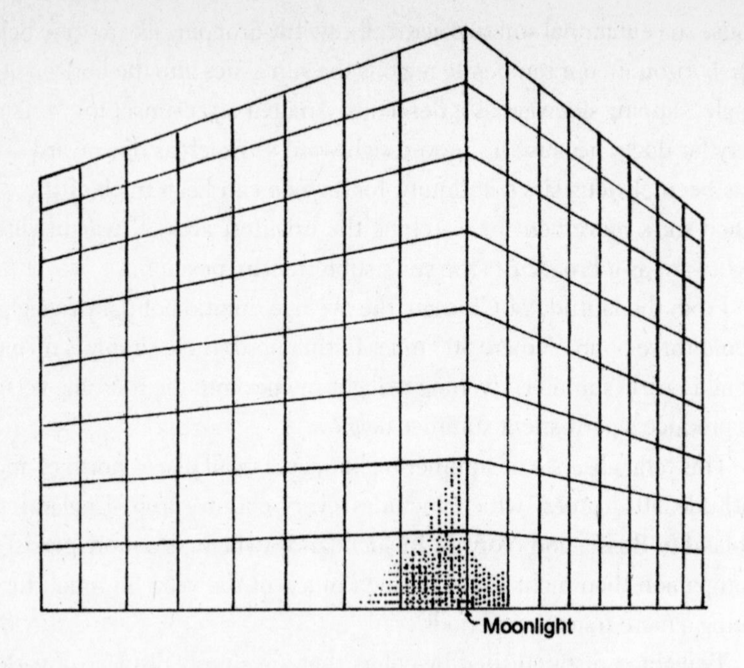

If the full moon's brightness is represented by the tiny cube at the near corner, sunlight's intensity is the whole stack, 77 by 77 by 77 moon cubes.

scatter away from the other colors. Thus robbed of the blue end of the spectrum, the remaining sunlight is heavy with the warmer pigments. That's why the setting sun appears orange.

But there's more to twilight than pretty colors. Diurnal animals including ourselves have evolved to take advantage of darkness' slow rate of onset. During this period our retinas gradually switch from their cone cells to the color-blind but highly sensitive rod-shaped sensors. The leisurely transition sets in motion other physiological effects as well. Pupil size increases, and photochemical changes occur in the retina, making it more sensitive.

In the process, human vision first loses colors at the ends of the spectrum—the blues, violets, and deep reds—while yellow and green linger much longer. That's why, during a stroll in late twilight when the light's intensity hovers near the threshold of color vision, we can still perceive the lime-yellow of a bicycle leaning on the porch even when the rest of the world has become a study in gray.

By the middle of astronomical twilight the change is nearly complete, and it's difficult to appreciate how dim the ambient illumination has become, for we never experience day and night "next to" each other. This explains why few people can accurately estimate the difference in intensity of moonlight and sunlight. When asked to place a value on the ratio, most guess that a scene illuminated by the full moon is anywhere from 20 to 500 times dimmer than the same vista in sunlight. No one approaches the true figure of 450,000 times dimmer.

In modern times, of course, the passage to night has been modified by artificial lights. Urban eyes are rarely asked to make a transition to anything resembling real darkness. Instead, the city dweller finds that day yields to a flux of constant illumination that roughly equals mid–nautical twilight. In major urban centers, full-darkness phenomena such as the northern lights can be seen only artificially, drawing appreciative gasps at planetariums. Seeing the Milky Way splayed across country skies, vacationing urbanites only slowly grasp what is before them. This beautiful panorama, this worm's eye view of our own galaxy's pancake-shaped structure, routinely enjoyed by our ancestors, has become a novelty in the nightlong synthetic twilight of the modern era.

The sun's actual position is shown by the dotted line. The image above the horizon is a ghost.

Real twilight—all of it—is fortunately still a daily delight in rural skies, and plays host to phenomena not seen at any other time. These events can be roughly divided into bright, early twilight phenomena such as antisolar rays, obvious even from cities, and the fainter curiosities seen best (or only) after the onset of astronomical twilight—like the legendary zodiacal light.

Let's stand outdoors, at least in our imaginations, and watch twilight's magic unfold. The experience, as we'll see, takes us up a stepladder of surprises.

We start with the orange sun sitting on the horizon. Surprise number one: It's not there!

Atmospheric refraction, which bends the image of the sun like a spoon distorted in a glass of water, produces a half-degree displacement at the horizon. By a wonderful coincidence, this just matches the sun's diameter. The air deftly deflects the sun's image upward from below the skyline. The orange ball hovering on the horizon is a ghost, an illusion. The effect is so dependable it's routinely figured into sunrise and sunset tables. The phantom sun gives us an extra four minutes of undeserved sunshine each day.

This daily bit of atmospherically induced mischief starts when the setting sun slows as it approaches the ground. It's really still moving at the same rate, thanks to the constancy of our rotating planet, but the sun's image increasingly appears higher than it really is. By the time it sits on the horizon, it's actually below the ground.

Here we pause for a moment, like millions of other romantics around the world, to watch the sunset—even if it is just a mirage. The process of the sun disk's transiting the horizon normally takes two minutes. In far northern areas, however, it sets at such a shallow angle that the sunset can be extended many times over, to as much as a full day near the poles!

During the two-minute sunset seen in northern temperate countries such as the United States and Japan, danger to the eyes has been greatly reduced because the tiger has been temporarily tamed; the shorter wavelengths of light including the dangerous ultraviolet have been filtered by the thick horizon atmosphere, which, you'll recall, is why the sun now appears orange or red.

If the air is dusty or humid, with distant cumulus clouds lurking beyond the western horizon, **crepuscular rays** will appear. These bright

Crepuscular rays: The pink glow on atmospheric dust or water vapor is interrupted by the shadows of distant cumulus clouds.

pinkish streamers, a commonly observed phenomenon of civil twilight, radiate from below the horizon so that their spikes (which are actually parallel) seem to converge at the sun's subterranean position. Who has not seen such magical rays from time to time, shortly after sunset, like a scene in a fairy tale?

In some parts of the world—South Pacific islands, for instance—crepuscular rays often cross the sky to meet at the point *opposite* the sun. This is a perspective effect, like the way parallel railroad tracks seem to converge. Such **antisolar rays** are prominent rather than

Looking east, we see avenues of sunlight hitting the upper atmosphere. Earth's shadow can be seen as a dusky band near the horizon, gradually rising as the sun sets.

subtle, and rate among nature's most curious spectacles. The spot at which they merge supplies a running commentary on the sun's departure, for its height in the eastern sky matches the sun's depth in the west.

A rarer phenomenon is the mind-boggling spectacle of a **sunset rainbow** flanked by those grand antisolar rays. A sunset rainbow is a stunning sight all by itself since (along with the even rarer sunrise version) it's the largest rainbow possible. The top of its arc reaches halfway up the sky, and the low sun provides a striking red background. The rainbow's arc is, of course, always centered opposite the sun, just where the antisolar rays radiate!

As the fiery ball continues to set, there is a moment when only a tiny orange morsel remains. This dazzling spot has gained worldwide attention, for it's the foundation of the **green flash.** Simply put, on rare occasions when atmospheric conditions cooperate, this final fragment of sun suddenly turns a vivid emerald-green. The color results from the refractive loss of colors at *both* ends of the spectrum. The blue end has already been lost through its long passage through the horizon air, but the red end can sometimes momentarily be filtered away by the right conditions. This leaves the middle, the green.

Put another way, the setting sun is actually a composite of numerous colored images, each slightly displaced from the others. Each component color is refracted a little differently, the red sun lowest, the blue highest—so that the "red" sun sets first, followed by the orange, the yellow, and finally the green disk, the uppermost. (The blue slice of sun would be last to set, but is usually absent. Still, there are photographs of the exceedingly rare "blue flash.")

I've seen the green flash 3 times, out of perhaps 150 occasions of searching for it. A clear cloudless horizon, preferably over the ocean, is the prime requirement. An odd and interesting apparition, the green flash lasts only for a few seconds.

Sunset over, the curtain rises on an esoteric but amazing wonder—the shadow of our planet! As the fading colors caress the western skyline, Earth's shadow rises at the opposite point of the sky—in the east. There it is, showing itself as a dark band hugging the horizon. This cu-

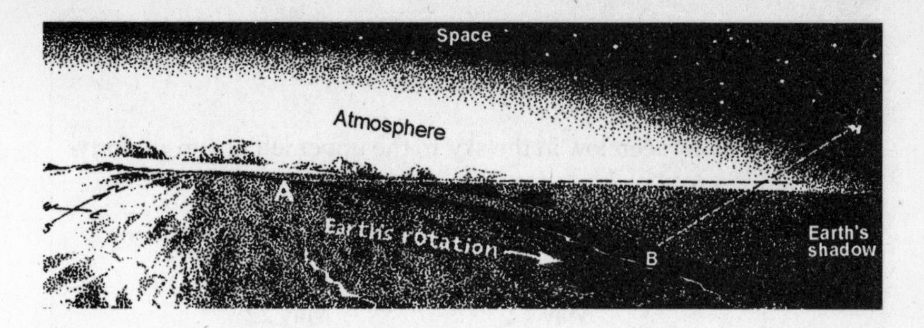

Standing at position A, which has just moved into twilight, you would be able to see Earth's shadow rising above the eastern horizon. Later, having moved to position B, you would no longer be able to sight cleanly along the juncture of light and shadow. But satellites in the sunlight above Earth's shadow are now visible.

rious apparition, easy to see if your low eastern sky is unobstructed, goes by an alluring name: **the twilight wedge.** Broadening as twilight deepens, the upper margin of this gray horizontal band grows more indistinct as your spot on the spinning Earth rotates further into darkness, for you're increasingly immersed within the shadow rather than sighting cleanly along its edge.

Above this boundary the sun still invisibly shines high overhead even after full darkness has descended on Earth below. This is the ideal stage lighting for the army of artificial satellites that begin to crisscross the sky at this time. (See the chapter "Satellite Season," page 179.)

Twilight also offers our lone chance for glimpsing the elusive planet Mercury. Mercury is usually the third or fourth brightest "star" of the heavens, yet despite such luminosity is rarely seen even by amateur astronomers. Flitting around the sun like a moth, it's usually lost in the solar glare, overwhelmed by the bright twilight that is its eternal home. But every spring there comes a two-week period when Mercury ventures out to the edge of its orbit just when its path lies most vertically above the sunset (see table, next page). At such times it isn't hard to see at all. Then its mercurial reputation as a challenging target becomes puzzling, for it floats conspicuously above the western horizon. Usually it reigns as the only bright "star" in that location.

Mercury in Evening Twilight

Mercury is easily seen low in the sky, to the upper left of sunset, forty-five minutes after sunset, within a few days of these dates.

1995:	Jan. 19	2001:	Jan. 28
	May 12		May 22
1996:	Jan. 2	2002:	Jan. 11
	Apr. 23		May 4
1997:	Apr. 6	2003:	Apr. 16
1998:	Mar. 20	2004:	Mar. 29
1999:	Mar. 3	2005:	Mar. 12
2000:	Feb. 15	2006:	Feb. 24
	June 9		June 20

Twilight is also the domain of Venus, the Evening Star, the brightest celestial luminary after the moon. Is it an airplane's landing light? A UFO? Here the problem is that it's *too* dazzling, so that people assume it can't be a natural object. Venus has its own chapter (see page 146), so suffice to say here that if the "star" hanging over the sunset is overwhelmingly the brightest in the heavens, you can be sure you've found Venus—which practically owns the twilight.

Near the end of astronomical twilight, the sky offers a final treasure: **the zodiacal light.** Appearing as a glowing band angling upward from the horizon, it's a specter reserved for locations free of light pollution. This pale apparition routinely materializes in tropical skies, where the zodiac, the path or plane of the planets, makes a steep angle with the skyline. In our temperate region it's noticeable only during those seasons when the zodiacal constellations sweep upward at their greatest angle: after sunset in early spring, and before sunrise between August and October.

Over the centuries many have been fooled into mistaking its glow for the first kiss of morning twilight: Khalil Gibran alluded to it as "the false dawn." Even more have been deceived by the evening version, misidentifying the eerie fluorescence long after sunset as a bit of lingering twilight.

The eerie zodiacal light, best seen after dusk in early spring

Its subtle radiance, broadest near the horizon and tapering upward, is the reflection of countless tiny points of sunlight from distant dust in the orbital path of the planets, millions of miles away.

Its soft creamy light—as if twilight were taking a last curtain call—heralds the imminent arrival of full darkness. Our planet's tireless turning then carries us relentlessly into the night, the zodiacal light's reluctant fading dousing the stage lights on the final echoes of the sun.

Transition period? Perhaps. But while human eyes may have evolved to take practical advantage of twilight's features, we probe its amazing phenomena with no thought of pragmatic value, but as a blueprint for better enjoying its magic. For who of us is not enraptured by its intoxicating light, and those "purple-colored curtains" which "mark the end of day"?

They knew, those Platters.

Autumn

The builders of the pyramids of Teotihuacán, Mexico, celebrated the zenith days, two days per year when the sun stood directly overhead at noon. The Pleiades (above end of road) *announced the first zenith day by setting at dawn. (See the chapter "The Lost Subaru." A Subaru parks appropriately in the foreground.)*

The Andromeda Connection

Galaxies form the largest structures in the universe.

Yet not long ago, nobody knew they existed. To the generation of the Roaring Twenties the cosmos was a single formation of stars; we lived inside of it and that was that. But photographs from the giant Mount Wilson telescope, built in 1917, eventually settled the matter: Those intriguing pinwheels, thought by many to be merely local solar systems, were revealed as island universes separated from ours by canyons of emptiness. Island universes! That beautiful and aptly august phrase eventually gave way to our present name for the immense cities of suns: galaxies.

At least 10 billion galaxies can be detected with today's telescopes, each galaxy home to hundreds of billions or even trillions of stars and probably an equally staggering array of planets, moons, and who-knows-what-else.

Yet the average educated person can name only one. A single renowned galaxy has so entered our vocabulary that it has become the galactic paradigm, the most famous by far. Andromeda.

One reason for Andromeda's celebrity is its fortunate overhead position above North America and Europe, where it's perfectly placed for observation each autumn. Probably another factor is its mellifluous name. After

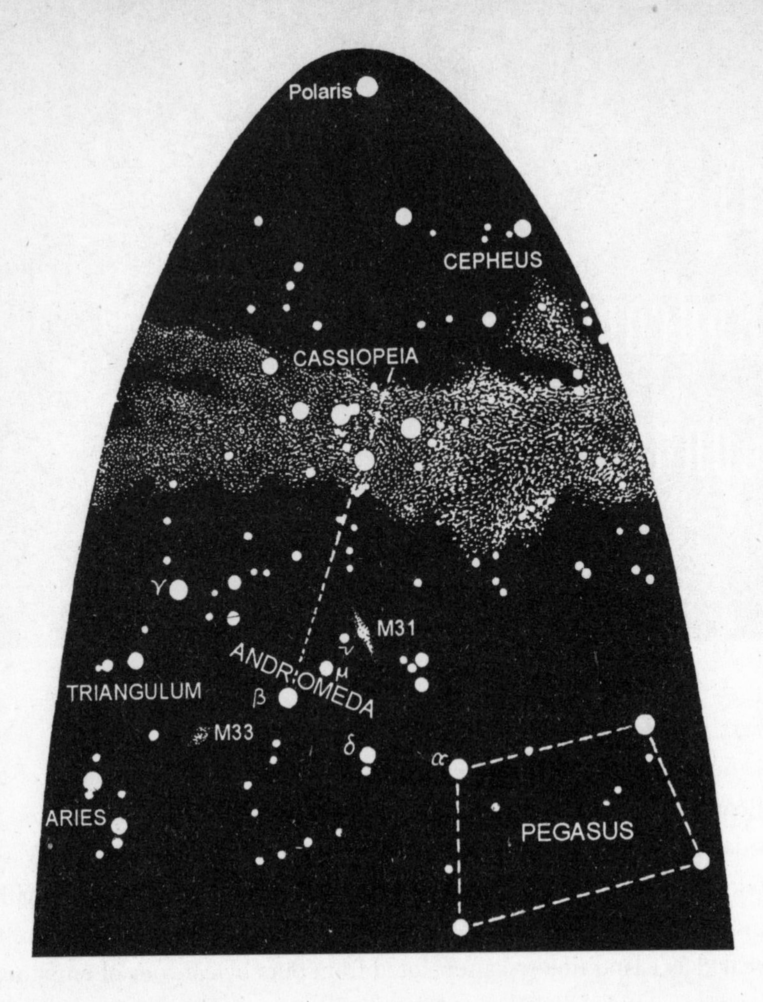

To locate the Andromeda Galaxy (M31), first find Beta Andromedae. Two approaches will work: (1) Work your way up to Beta (β) from Alpha (α) at the corner of the great square of Pegasus, or (2) use the larger triangle of Cassiopeia as an arrowhead to point to Beta. From Beta, Mu (μ) and Nu (ν) lead to M31. With binoculars you can also find the much fainter galaxy M33, the third spiral in our Local Group.

all, the galaxy M87 is bigger, M82 more violent, Centaurus A more mysterious. But none have the euphonious appeal of Andromeda, a name too lovely to forget.

It really didn't have much competition. Only a dozen or so of the most extraordinary galaxies possess proper names; millions of others are merely cataloged by numbers or go as unlabeled as autumn's falling leaves. There wasn't even any special reason for the Andromeda Galaxy to be named after the maltreated woman of Greek mythology. It was just a fluke of real estate; the galaxy merely took the name of the constellation in which it lies.

Senior citizens and older textbooks still call it by its former title, used for centuries: the Andromeda Nebula. Countless generations saw it only as an intriguing cloud of unknown nature, though many sleeves were insistently tugged by its perplexing paradoxes. For example, through a spectroscope it displays the signatures of countless stars, yet no star could be seen there. Only with the staggering revelations of its true size and numbing distance could we fathom that, like faraway rain whose drops seem merged into a blurry sheet, its numberless suns are reduced to fog by the millions of light-years separating us.

Andromeda is more than bright and pretty. It's also the very nearest spiral galaxy, and the largest in this region of the universe. This itself is curious—that the nearest is the largest as well. It's the kind of coincidence that makes scientists suspicious of their data. Recent studies, however, confirm Andromeda's imposing size: We'd have to venture 50 million light-years to find anything bigger.

But its greatest draw may simply be that it's there—visible to the unaided eye. Our naked-eye view of *all* external galaxies reveals Andromeda, period. All else is foreground. Even the lavish star fields that spill across autumn skies like blowing sand, so seemingly infinite, are next-door neighbors compared to the ghostly glow lying a thousand times farther in the distance. Looking like a faint fragment of fog, utterly camouflaged in bright city skies, Andromeda challenges our minds to grasp the contradictory concept that the nearest galaxy is also the farthest light detectable to the naked eye.

How distant is this oval wisp? Astronauts traveling fast enough to touch the moon in just three days would need 500 billion years to reach Andromeda. That's dozens of times longer than the universe has existed. Like snowflakes blown against a window, the night's stars are mere foreground specks bearing no spatial connection to that enormous object looming in the distance.

It's big, all right. Otherwise, how could something so remote take up 4 degrees of our sky? Even its brighter central section, the part seen by the

M31—the Andromeda Galaxy—with its two companion galaxies, NGC 205 (top) *and M32, which is so close to the far edge that it has distorted Andromeda's spiral.*

unaided eye as a glowing oval, appears far larger than the moon. Its enormity might be grasped by this comparison: Moonlight reaches our eyes after traveling less than two seconds, but Andromeda's frozen portrait arrives after hurtling through space for 2 million years.

At the very heart of Andromeda, astronomers believe, lies an immense black hole. Weighing as much as a million suns (or 300 *billion* planet Earths) it is the glue that holds the core of the galaxy in place. Around this invisible, ultradense region whirl a trillion suns and probably an equal number of planets, as if paying obeisance to a vast unseen temple. Like a tempest surrounding the eye of a hurricane, more stars orbit this mysterious spot than could be counted, nonstop, in ten thousand years.

In 1993 the Hubble Space Telescope showed that a bright cluster of stars lies a few light-years from Andromeda's exact midpoint, while a smaller cluster sits at dead center, both presumably orbiting the black hole. This dual cluster is unique among galactic cores and remains a target of continued exploration.

Imagine a spaceship that could whiz round-trip from California to New York and back thirty times in a single second. Now, using this speed—the speed of light—fly away from the center of Andromeda. At such a pace you'd endure fifteen thousand years of travel before you'd emerge from the nuclear bulge at that galaxy's heart. Then you'd start to explore its awesome spiral arms, stretched like curving pinwheels sprinkled with untold spheres of nuclear flame. This is the adolescent zone, the region where young blue suns proclaim their ephemeral existence with blinding diamond splendor. Another fifty thousand years of outward travel would carry you through the outer arms, a more spacious region where, perhaps, the riot of local activity would thin enough to let your attention be caught by our own Milky Way Galaxy floating in the distance.

Beginners assume the heavens' nearest and brightest spiral galaxy must be simply awesome through a telescope. Unfortunately, not so. Galaxies are usually disappointing to the visual observer because their myriad stars merge into an unexciting smudge. Then, too, there's a certain amount of "show biz" behind the presentation of celestial objects. Meaning that people's expectations play an important role in how an image is perceived. When you invite Uncle Larry and Aunt Jane to look through the telescope, they'll not have heard of M13 and thus will be surprised and delighted by the dazzling stars of that globular cluster. But point the telescope at Mars or Andromeda, whose names are familiar to everyone (but whose subtle detail demands an experienced observer's eye), and you'll get only a polite response—assuming Uncle Larry has suddenly become tactful in his old age.

Andromeda is blurry through even the finest instruments, since its pointillistic glory has been reduced to a foggy patch by the awesome emptiness between us. But if you still want to check it out for yourself, use your telescope's minimum magnification. The biggest problem when observing Andromeda is obtaining a power sufficiently low to allow the whole galaxy to fit in the field of view. The task is impossible through most telescopes. Andromeda's just too big: Only a small, unimpressively soupy section shows at a time.

Larger instruments do brighten the image, but no telescope has ever visually resolved the stars in any galaxy. At our observatory we rarely show Andromeda during public viewing nights because it's anticlimactic. (By

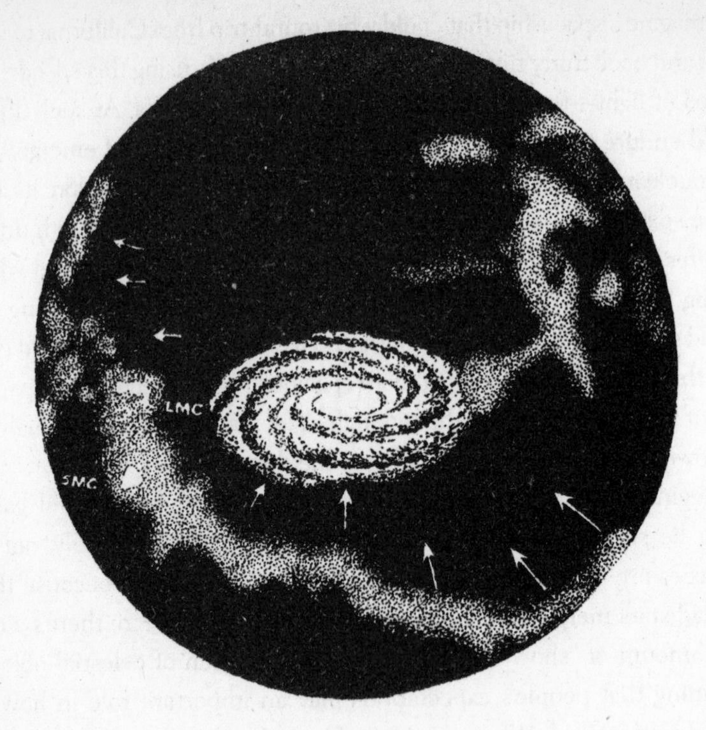

The Large Magellanic Cloud (LMC) and Small Magellanic Cloud (SMC) are galaxies immersed in a huge gas cloud that passes under the Milky Way and may be in orbit around us. Parts of the cloud are rapidly falling toward our galaxy (arrows).

contrast, the galaxy M51 on the opposite side of the sky, while a hundred times fainter, is a snappier sight thanks to its prominent spiral arms.)

Telescopes do reveal two cloudy patches hovering alongside Andromeda like attending fairies. These are dwarf companion galaxies with *only* a few billion suns in each, just as we ourselves are accompanied by the two Magellanic clouds. Such small, irregular galaxies fail to excite us visually; they don't stir us like the lovely spirals. But galaxies of the less glamorous type are more common in the universe. They're like the cruisers, tankers, and other smaller ships floating alongside the aircraft carrier.

The Magellanic clouds (so nearby that many astronomers believe they lie inside the tenuous halo of our own Milky Way) can be seen only from tropical or southern climes. From Europe and from the United States and Canada they forever hide below the southern horizon, blocked by the bulge of Earth. Looking indeed like small detached fragments of cloud, and boasting no structure beyond that of a blob of oatmeal tossed onto the ceiling by a bored toddler, neither galaxy is impressive. But because they are the nearest galaxies to our own, these midgets force the qualification of describing Andromeda as the closest *spiral* galaxy, or closest *major* galaxy, or the only galaxy visible from our region of Earth. The many writers who simply label Andromeda "the nearest galaxy" risk a lawsuit from some Magellanic trade organization.

Binoculars are superb for enhancing Andromeda's presence, especially those with 35-millimeter lenses or larger. (That's the second number in binoculars' specifications, as in 7 × 35. The first number is the magnification. See Appendix 3.) You can find our sister galaxy through any pair by sweeping the overhead sky at 9 P.M. in the middle of November; its oval patch stands out nicely.

Long-exposure photography is what's really needed for an impressive galactic image, since film can accumulate faint detail while the eye cannot. But here as well, Andromeda's no pin-up model because it's tilted only 13 degrees away from edge-on to us, disguising its spiral arms no matter what we do. Their positions are suggested by black dust lanes, the galaxy's most imposing features.

By sheer coincidence, the nearly edgewise appearance of Andromeda matches the view any Andromedans would have of us. The fact that Andromeda floats near the Milky Way in our night sky automatically tells us that from there, we, too, are oriented sideways. A shame, since the sight of such a majestic nearby pinwheel would be awesome if either of us had been presented face-on to the other, instead of edge-on.

Still, things could be worse. We are presently on the side of our rotating Milky Way Galaxy that is closest to Andromeda. In another 120 million years, when we've circled to the other side, Andromeda will probably vanish behind the bulge at our galaxy's center, like an onlooker hidden by the center of a carousel.

It's a safe bet that intelligent life gazes upward from numerous Andromedan planets, curious about our smudgy galaxy floating in their sky. If only one planetary system in a *billion* contains creatures with intelligence, then thousands of such worlds are hidden anonymously in the silky folds of Andromeda's spirals. Perhaps they even realize, as we do, that our two galaxies are approaching each other. With the separation decreasing by nearly 50 miles each second, there's a reasonable chance we'll collide in 4 billion to 6 billion years. But no problem: Colliding galaxies are all smoke and mirrors, passing through each other without harm, since their constituent stars are too far apart for individual contact. Some, however, would be yanked by gravitational forces from one galaxy to the other, offering us the attractive prospect that we may someday switch allegiance and join Andromeda.

We can only hope so. The return address on our envelopes has read "Milky Way" for far too long. That we've been stuck with the most ridiculous of all galactic names is an indignity we've borne with quiet patience. The opportunity to jump ship, especially to a name like Andromeda, is an event worth waiting for. It should be marked on the calendar.

If we don't collide, we'll still hang out together. Like its mythological namesake, Andromeda is chained to the rock of our own gravity and will always linger next door. While virtually every other galaxy has joined the "let's fly away from the Milky Way" bandwagon, Andromeda is one of just

Andromeda, as it might appear from a planet in the Small Magellanic Cloud, hangs just right of center in this illustration. The spiral dominating the sky above is our galaxy. Its nucleus is less than a tenth as distant as Andromeda's. Coincidentally our sun is situated directly above Andromeda and left of our galactic center.

two dozen that loiter too close to participate in the universe's size-increase program. In a cosmos where the expansion-indicating redshift is as commonplace as tuna salad, Andromeda, by approaching us, displays the rare **blue shift.**

So whatever the fate of the universe, we'll occupy adjacent seats for the show. Reason enough to take a peek as it passes high overhead these nights. But if more motivation is needed, you might simply recall that when looking at Andromeda, you're gazing at the farthest thing your eyes can reach.

The Once and Future Past

How strange and intriguing, how tangible and yet dreamlike is the dimension of time! Science fiction often weaves through the temporal empire, but we can enjoy real-life encounters with the fickleness of time whenever we gaze into the night sky.

Want to see the past? Watch time slow down? Experience clocks ticking at different rates? These are routine occurrences in the alleyways of the night. And with daylight saving time ending, forcing autumn's ritual of clock tinkering, this is a good season to look at time itself.

Let's begin with the fact that we never really see what *is,* but what *was.* Images, although traveling at light's impressive 186,282-mile-per-second speed, take time to reach our eyes. Here on Earth you can quickly judge the lag because of the fabulous coincidence that light requires a nanosecond (a billionth of a second) to travel one foot of distance. When you look at anything—your neighbors, say—you're seeing them not as they are, but as they were as many billionths of a second in the past as they are feet away from you. Standing 40 feet from your eyes, the neighbor with the lawnmower is seen performing the actions that occurred 40 nanoseconds ago.

This discrepancy, so easy to calculate, is without practical consequence until we raise our eyes upward. For the moment we consider the denizens

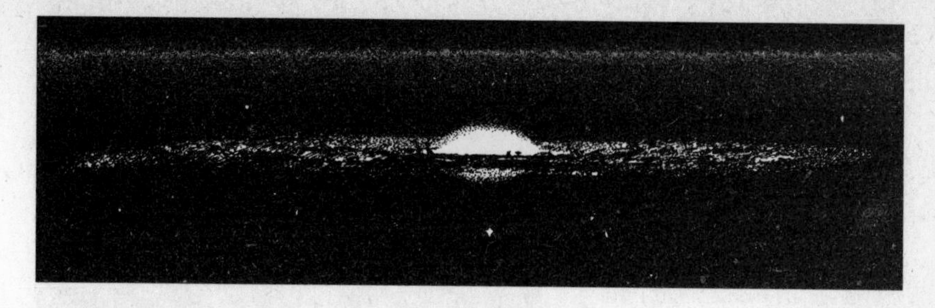

The light arriving from the beautiful edgewise spiral galaxy NGC 4565 started its journey as the dinosaurs were enjoying their last supper.

of the cosmos, we are suddenly reading yesterday's newspaper. The image of even the nearest planet, Venus, requires several minutes to arrive at our eyes, while the nearest stars are years away at light speed. Brilliant and relatively nearby Vega, lingering in the west in early autumn, has its image arrive some quarter century after it started out. If aliens living on a Vegan planet were observing you this instant through some kind of supertelescope, they'd now be watching you do whatever you did in the late sixties. (Those wild years still live on in the annals of the Galaxy!)

There is no possible method by which faraway creatures can have current news of Earth, nor we of them. This is why astronomers are so intrigued by distant objects. They're dim and blurry but open up a scrapbook of what things were like in a bygone era.

We can, for example, compare the appearance of remote galaxies with our own, to see if there has been celestial evolution through the aeons. This is not mere theory or abstraction. Space equals time, and the practical consequences can be stunning.

Take quasars. Archives of mystery, these tiny objects were discovered in the 1960's and found to be the brightest, most distant, and fastest-moving bodies ever known. At first all manner of exotic theories emerged; some thought they might be material gushing from a white hole, after having fallen into a black hole somewhere else—in short, matter tunneling through space or time.

The explanation proved to be simpler, if just as profound. Quasars are the unimaginably brilliant cores of young galaxies, as if numerous supernova explosions are going off en masse. But something else makes them unique as well: None live in our part of the universe. They're seen *only* far

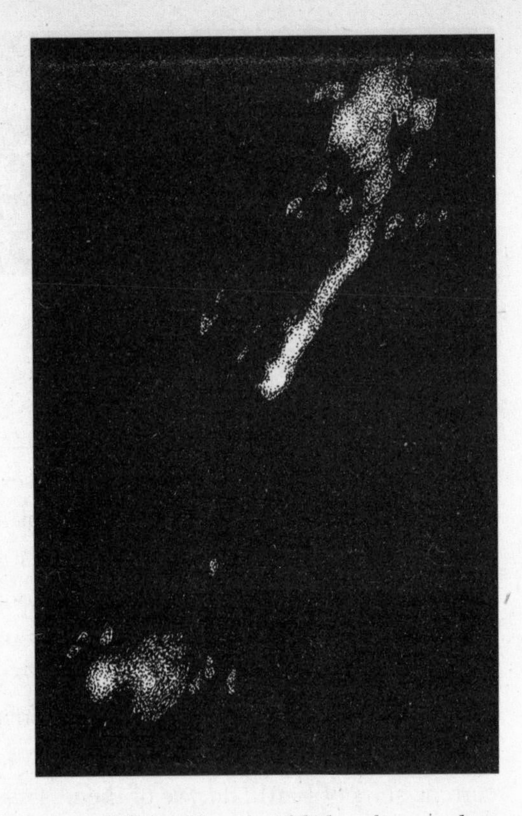

Fifteen billion light-years distant, quasar 4C 32.69 appears as a point of light in visual frequencies, but in radio frequencies (here six-centimeter wavelength) the point (center) is flanked by two lobes and a jet, indicating the active core of an invisible galaxy. The extraordinarily long time required for its light to arrive means we are viewing events that unfolded soon after the universe's birth.

away, billions of light-years distant. Why? Why should they be avoiding our region of the cosmos?

The answer is elegant and instructive. Quasars are a phenomenon of the young universe. They do not exist anymore, anywhere. They display the kind of ultraviolent, frenzied activity that galactic nuclei *used* to experience. Perhaps our own galaxy and nearby Andromeda were quasars in their youth. And they'd still look that way to creatures in the distant reaches of the cosmos, for whom the light from the infant Milky Way is just now arriving.

We see only those quasars far enough away to have images still en route from ancient empires. Quasars are not avoiding our space; *they're avoiding our time.*

Most people understand this principle. They often repeat the myth that many of the night's stars do not even exist; we see only their images. Though false, this idea shows a correct perception of the disparity between appearance and reality, between image and actuality. (The rea-

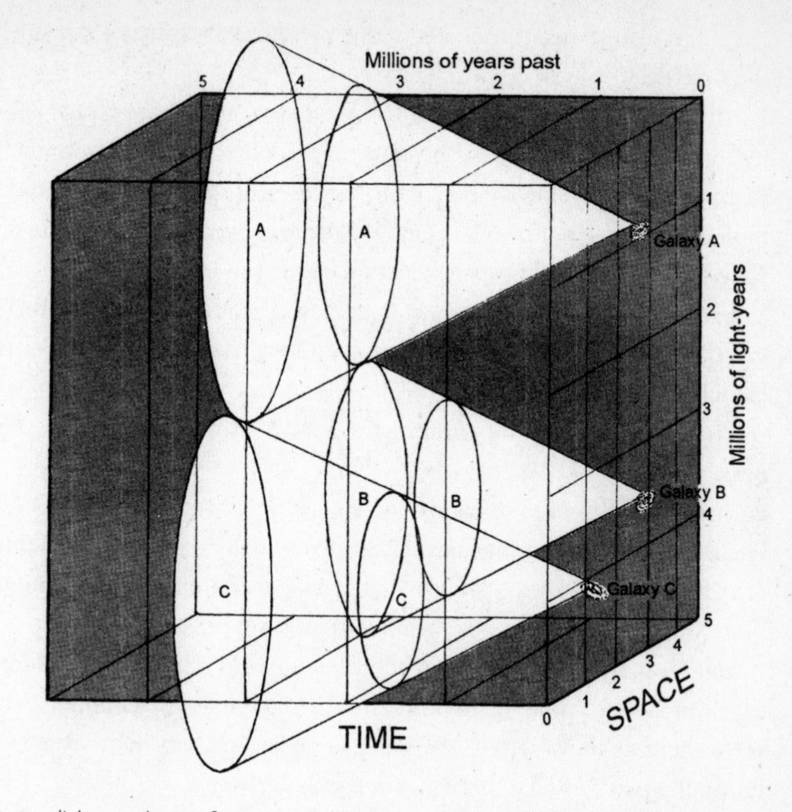

Because light travels at a finite speed, seeing across space is also seeing back in time. To show this, we have simplified space into the two-dimensional grid at the right and used the third dimension to represent time. Each observer can see only whatever is within the cone expanding backward in time from her present position. She sees another galaxy only as it was at the time in the past where two cones meet. For example, residents of Galaxies A and B view each other 3 million light-years in the past. B and C get news that's 2 million years old.

son it's untrue is that a star typically lives for billions of years. The odds of its dying in the brief few hundred—or at most few thousand—years during which its image travels to Earth are minuscule. Multiply those odds by the six thousand stars visible to the naked eye and it's still hundreds to one against even a single star's no longer being there. But the principle is sound.)

The present moment occurs only in your immediate vicinity. And that, while intriguing, is not a mysterious notion at all. It is simple common sense, a consequence of light's finite speed, and has nothing to do with the weird distortions of time and space predicted by Einstein's relativity, the time dilation we'll come to shortly.

Time can also appear to run slowly, thanks again to light's finite velocity through space. When we observe an object racing away from us, its gestures unfold in slow motion, while any creatures living there would simultaneously see our own movements occurring at an artificially sluggish pace. A simple thought experiment can clarify this strange effect.

Imagine blasting off in a rocket and telescopically observing the receding Earth through a rear-facing window. Your friends, standing on the launching pad to see you off, applaud your courage and perhaps the successful attainment of the government grant that made the whole thing possible.

With each clap of their hands, you are a little farther from Earth, so the image of the next clap has farther to travel. It thus arrives a bit behind schedule, and the result is a view of people applauding in slow motion! From your perspective it's been a lukewarm send-off.

You stop the rocket upon reaching your destination and suddenly earthly images resume their normal speed. Of course, you continue to observe them as they were in the past, since if you're six light-years from Earth, the pictures take that long to reach you.

During the trip back home, your forward-facing telescope sees earthly events unfolding in speeded-up Keystone Kops fashion, because you now intercept each image ahead of the time it would reach a stationary ob-

On a spaceship capable of two-thirds the speed of light, you could observe Earth evolving through only three years during your outward trip of nine years. On returning, fifteen years of Earth history will be crammed into your next nine years of observation.

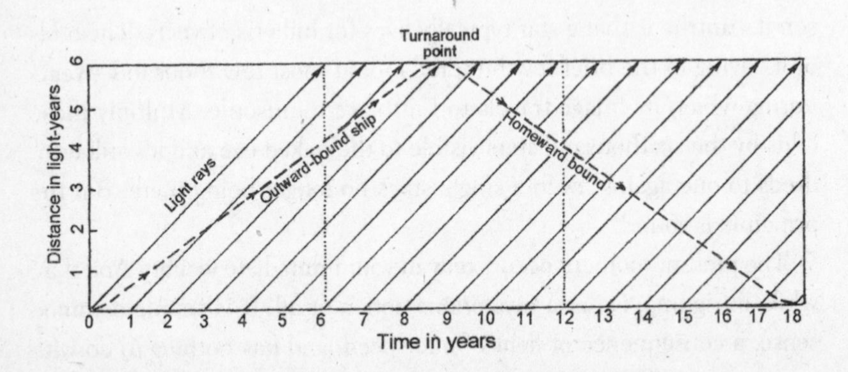

server. This comedic high-speed digest of the news you've missed during your absence continues until landing, when image and reality coincide, and events once more unfold in "real time."

Again, there's nothing mystical about this; it's simple logic. But it does mean there are star systems where, using a supertelescope, aliens would observe life on Earth not just as it was in the past, but in slow motion as well. Can you imagine having to watch congressional speeches performed at a snail's pace? They'd abandon any plans to visit.

Such phenomena can be put to practical use. One way is to observe galaxies at the distant edges of the universe, and determine how fast they're expanding away from one another. Then we compare those speeds with the ones in our own neighborhood, and see how much the expansion of the universe has slowed—since the difference between *there and here* really translates into *then and now.* Extrapolating into the future allows us to predict whether the cosmos will expand forever, or if instead the deceleration is enough to bring us to a future standstill and eventual collapse.

Some astronomers are actively working on exactly this problem; increasingly accurate measurements are expected to solve this great puzzle, an issue that affects every planet in every galaxy.

Another time-related problem was solved as recently as 1993. By comparing distant galactic clusters with those in our more immediate vicinity, astronomers found that those stellar empires *evolve.* Clusters so far away they're only half the age of our own Local Group of galaxies contain more *spiral* members, and many more galaxies in collision and near collision. Apparently, back when the cosmos was a more crowded place, galaxies often bumped and merged with each other. This resulted in the modern neighborhoods composed largely of elliptical members, which we now assume are the product of grand galactic collisions between spirals.

More astounding than these changes in time's appearance, and harder to understand, is the way time warps and slows on its own, the **time dilation** aspect of Einstein's special relativity theory.

In everyday life, "time running incorrectly" can be experienced simply by having one's watch repaired in Central America. Subtler changes are produced whenever we walk up or down a few flights of stairs; the slight alteration in gravity caused by varying our distance from Earth's center is enough to produce measurable (at least with atomic clocks) changes in the flow of time.

The two ways we can throw a monkey wrench into the passage of time, then, are by traveling faster and by vacationing in a place of powerful gravity. The kind of gravitational field strong enough to do the job effectively would not be compatible with life. But high speed is another matter, since people can safely attain it in gradual increments.

Time slows appreciably only if you get close to the speed of light. At half light speed, for instance, the change amounts to just 20 percent. That is, your clocks advance, and your body ages, 20 percent slower than for people back on Earth. But attain 184,500 miles per second—99 percent of light speed—and the change is 700 percent.

At that velocity, a day elapses for you while a week passes on Earth. If you traveled at this rate for, say, a decade, nothing would seem amiss on your spacecraft. The passage of time would appear normal, and you and the other crew members would look ten years older at the end of the trip. But upon landing on Earth, you'd discover that seventy years had passed. Probably no one you left behind would still be alive.

An obvious question arises: Whose time was right, and whose changed? But the answer—and this is the whole point—is that there simply is no right or wrong passage of time. Simply, ten years passed for you while *at the same time* seventy years elapsed on Earth!

Time is thus a flowing river that varies its rate depending on local circumstances. It does not exist as an absolute. There are places in the universe where a million years elapses while a mere second *simultaneously* passes for us—and vice versa.

This seems alien only because we grew up in an environment where everybody and everything incestuously share the same motion and gravitational field. But there are sites in the night sky where we can directly confront realms of disparate time.

Anyone with a telescope can gaze at several of the fantastically compact white dwarf stars whose strong gravity slows the matrix of time. The most famous example is the white dwarf companion to Sirius (which has its own chapter, "Two-Dog Night," page 71), but the easiest to observe is the brighter of the two companions to the winter star Omicron 2 Eridani.

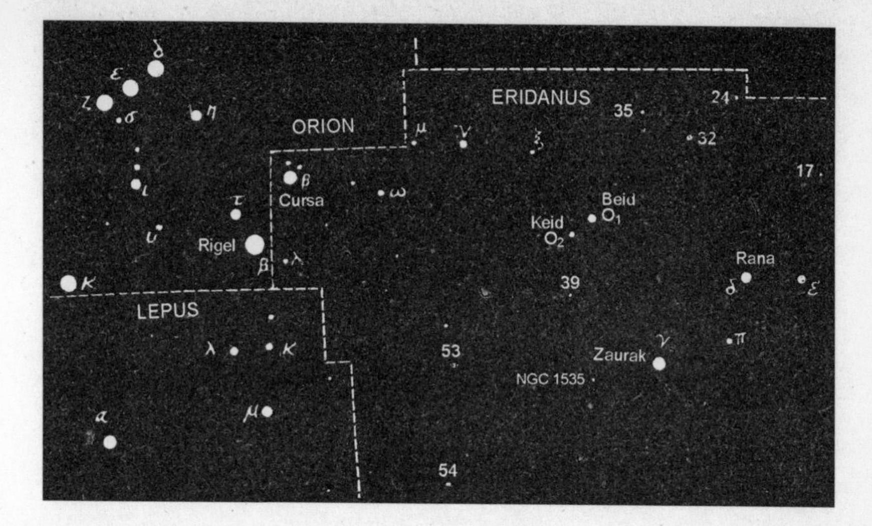

Omicron 2 in Eridanus (Keid) lies west (right) of Cursa on a line with Omega (ω) and Delta (δ). It is a triple star system, the white dwarf being the second brightest. A small telescope easily picks it out.

A ten-inch telescope can do even better. With it you can observe the sky's brightest quasar, which also wins kudos as the most distant object seen with common amateur equipment. Its name is 3C273, and it sits about 5 degrees northwest of the star Gamma Virginis.

Quasar 3C273 lies some 2 billion light-years from us and is zipping away at 30,000 miles per second. This represents about a sixth the speed of light, enough so that 3C273's clocks run so differently from ours that you'd catch the discrepancy in just a few minutes, with nothing more sophisticated than an egg timer—if you had a telescope powerful enough to read the clocks there.

Moreover, any creatures there would now be observing Earth as it was billions of years in the past, long before the epoch of the dinosaurs. And this is the *nearest* quasar. The world's largest observatories have photographed quasars as far away as 10 billion light-years, where no telescope, no matter how powerful, could see our solar system—since the image of our sun's birth won't reach them for another 5 billion years. If they're peering in our direction at this moment, they see nothing but the nebula that draped our neighborhood before Earth was born. If we had known they'd be looking, we could have erected a giant billboard saying WATCH THIS SPACE to keep them interested.

The curious fickleness of time cries out for exploitation. It seems likely that we'll someday launch interstellar astronauts at 99 percent of light speed. They might complete their round-trip odyssey thirty years later by their own clocks—two centuries later Earth time—to share their escapades with the great-great-grandchildren of the people who sent them off.

Perhaps. For now, the shimmering of the night's stars seems to symbolize the Sisyphean futility of tethering time to our static assumptions. Flowing its own way, like a rapids roaring here but quietly pooling there, time runs its course while our calendars and clocks indulge themselves in the earthly illusion of constancy.

The Lost Subaru

No obstetrician attended the birth of the Pleiades, 20 million years ago, and it's just as well. The fierce blue-white nuclear heat emitted by the septuplets sterilized everything for billions of cubic miles in all directions. As additional suns awakened from the dazzling and dangerous nursery, the newborns materialized like a distant sunrise in the skies of Earth 410 light-years away.

Today this coterie of stars stand as an oddly unique sight, a single bright, conspicuously dense cluster in the autumn heavens. The popular name of this stellar hatchery? You probably already know it (and even own the ideal instrument for observing it in all its glory). It's the Seven Sisters.

The vivid group of huddled stars often confuses beginners who mistake it for the Little Dipper since it *is* little and *does* look a bit like a dipper. Experienced sky watchers snicker at such an error, for the real dippers, far larger, remain glued to the north, the one sky-neighborhood the Pleiades never visit in their nightly rounds. Nowhere else in the heavens can one find anything like them. Visible even in the bright skies over cities, the beelike swarm of stars is prominently positioned for 80 percent of the world's population. Perhaps that's one reason cultures throughout the ages have given this asterism special attention. (Not a separate constellation, the Pleiades are just a small part of Taurus, although nobody's ever figured out what bullish body part they're supposed to portray.)

Floating in the eastern sky as darkness falls, they seem benignly intriguing. But that wasn't always the case. In ancient times the Pleiades had a strange, sinister reputation. Such medieval rituals as the pagan Black Sabbath and All Hallows' Eve (which evolved into our own Halloween) were set to occur when the Pleiades culminated—reached their highest point at midnight. W. T. Olcott's classic *Field Guide to the Skies* speculates that the rituals could have originated as a sort of commemoration of some ancient catastrophe that resulted in great loss of life. Robert Burnham, in his superb *Celestial Handbook,* suggests that all this may be linked to the Atlantis myth, itself perhaps evolved from the awesome eruption of the Santorin volcano in 1450 B.C., which devastated the Minoan civilization on nearby Crete. If indeed that disaster occurred at the same time of year as the Pleiades reached their midnight culmination, the cluster may have become a sort of "memorial for the dead."

Nowadays such dark undertones have been forgotten, and trick-or-treaters make their rounds thoroughly oblivious to the Seven Sisters overhead, or the possible ancient and sinister association between the stars and the revelry.

The Pleiades had a strange importance to civilizations throughout time and around the world. In Egypt they were revered as one of the forms of the goddess Isis. In ancient Persia the date on which they reached their highest midnight ascendancy was marked with ceremony. In the Mayan and Aztec cultures, this same yearly occasion had a forbidding undertone and was given tremendous importance—with at least one city's streets and pyramid aligned with the setting of the Pleiades.

In Japan their ancient name is Subaru. The six companies that merged to produce automobiles in 1953 under this name continue to place a crude star map of the Pleiades on each of their cars. The logo recently changed, with one star now brighter and more separated from the others than before, perhaps revealing some cryptic corporate ascendancy.

A glance upward shows that one Pleiad *is* brighter than the others: Alcyone (al-SIGH-o-knee), a pastel-blue giant that emits the light of a thousand suns. But all the siblings have easy names to remember if you're a mythology buff: They're Atlas and his seven daughters.

But why *seven* sisters? That's the real mystery. After all, normal eyesight can readily count only six, the same number found on the Subaru insignia.

If you *can* see a seventh, then you should be able to see an eighth and ninth as well. How many you can perceive tells as much about the purity of your sky as the state of your vision. With good eyes in a rural setting, nine are a cinch, eleven aren't too difficult, and as many as sixteen have been reported.

The real thrill comes when the proper instrument is pointed their way. Not a giant telescope; that would be a mistake. Far better is a simple pair of binoculars, because low power and wide field are the ticket. The minimum magnification on most telescopes is way too high, preventing the entire cluster from fitting in the field of view along with some encircling black space for an aesthetic backdrop.

Beginners often gasp when first seeing the Pleiades through binoculars. Suddenly dozens spring into visibility, their blue-white color becoming obvious as well. After you've spent a few minutes counting them with the naked eye and thereby learning the configuration of the main stars, ordinary field glasses provide a wondrous follow-up, instantly demonstrating the surprising astronomical value of a common household pos-

The Pleiades through binoculars

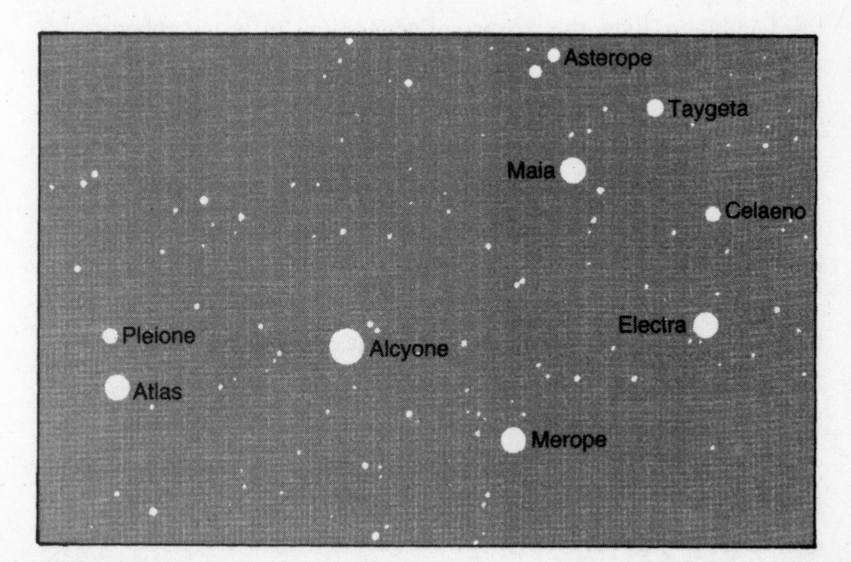

session that often lies unused and undervalued. Indeed, if your binoculars could offer you their single most amazing celestial vista, the stunning and dramatic transformation in the appearance of the Seven Sisters would probably be the winning entry.

The Pleiades once again illustrate that high power, so mistakenly prized by beginners, is often more of an impediment than a blessing. To display a pleasing image, each celestial target has an optimum range of magnification. When those parameters are overstepped, the result is a sharp aesthetic diminishment. In the case of the Pleiades, no more than 30 × should be used. The 7 × to 10 × offered by most binoculars is close to ideal.

But again, why *seven* sisters when the average eye sees six? Why have civilizations as disparate as the ancient Greeks, Australian aborigines, and Japanese all possessed legends of the "lost Pleiad" that have persisted through the centuries? Even two thousand years ago, the Greek poet Aratus wrote: ". . . their number seven, though the myths often say . . . that one has passed away."

Has a Pleiad changed in the astronomically brief instant that has elapsed since the epoch of the Parthenon? At first it seems unlikely, since the lives of stars play out on a vastly slower time frame than the mere millennia since humans first linked the sisters with the number 7. To our eyes, deep-space objects such as the Pleiades remain frozen in posed, fixed grandeur. Even the passage of generations is far too ephemeral to observe their stately gestures. But some intriguing evidence suggests that here may be an exception; in this corner of the museum the statue may have moved when we weren't looking.

First clue: They are, as binoculars reveal, blue, a color that indicates stellar youth. Young, hot, giant stars are often unstable, gobbling up their nuclear fuel in an adolescent frenzy that frequently produces instability.

One in particular, Pleione, shows spectroscopic evidence that it may flicker like a wind-blown candle. The behavior of this star, emitting a peculiar shell spectrum and slightly varying its light even now, may be responsible for the cryptic comings and goings of the "lost Pleiad." Another youthful mannerism is the dizzying rotation of the Pleiades stars. Pleione spins more than one hundred times faster than the sun, its surface flashing along at a fantastic 200 miles per second. Clouds of ultrahot vapor shoot from its whirling equator to produce ever larger, spiraling shells of gas that surround the star. This strange and turbulent environment is a likely breeding ground for brightness fluctuations.

It is not easy to tell, however, just when the "lost Pleiad" faded from view. The words of ancient Greek observers suggest that even by then the seventh Pleiad had dimmed since an earlier epoch. Perhaps, since the number 7 had more traditional significance than 6, this handful of stars was called the Seven Sisters in utter disregard of the true number!

In any event, the entire sapphire neighborhood of the Pleiades, with its delicate tendrils of smoky gas and madly whirling suns, is an astonishing and powerful spectacle as well as a fascinating astrophysical laboratory. A nineteenth-century astronomy text called the Pleiades "the meeting place in the skies of mythology and science."

Astrophotography helps us probe this section of space. Photos reveal a wispy blue nebula wreathing the entire neighborhood, glowing like phosphorescent frost on a window pane. Thickest just south of Merope, this vast cloud of dusty gas is the celestial embryonic fluid from which all the Pleiades were born. Stars always evolve from nebulas, but the Pleiades were created so recently that their cosmic afterbirth still lingers.

Portrait of the Pleiades. The stars are embedded in a cloud of dust that scatters blue light in our direction the way our own atmosphere scatters sunlight to create our blue sky.

They're in their infancy even today: our own sun has been around 250 times longer. The dinosaurs gazed unconcernedly into a sky empty of the sisters, which sprang into view just before we ourselves did. And, since such massive stars die young, the Pleiades will be long gone when most of the Galaxy's stars are still enjoying middle age. Toddling gracefully across November's chilly skies, the newborn sisters are only for now.

Great Balls of Fire

The circle. Through the ages it was considered nature's "perfect shape." The Church *insisted* celestial objects must travel in circular orbits, since God would surely never allow His creation to meander in anything but flawless paths. Such reasoning was itself circular: It somehow didn't occur to the theologians of the day that all shapes might be "perfect." In any case, the circle *is* nature's favorite shape and, arguably, its most amazing. And in the distant depths of the night, bizarre circles, rings, and spheres with astonishing properties have been discovered.

Let's explore a selection of cosmic circles that are part of the autumn sky. We'll start modestly, close to home, and spiral outward to the new-found globes of nothingness that enigmatically constitute the essence of the universe itself.

First, though, why *are* spheres so common in space? Why do sun, moon, planets, and stars not sometimes assume the guise of cubes or polyhedrons or diamonds, like the lovely gemstones that they are?

The answer is wonderfully simple. A celestial body starts out gaseous or molten, and is thus easily malleable into any shape. All its atoms attract each other by simple gravity, so it pulls itself inward to the most compact figure possible—which is always a globe.

You discovered in childhood that a sphere has the smallest surface of any form. When you played with clay, you could pattycake it into a thin piece with an enormous surface, or could roll it into a satisfying little ball between your palms. A ball was always the tiniest you could make it, with the least amount of surface area. If you were running low on paint, a final shape closest to a sphere would require the least amount.

Only objects with too little gravity to do the job—small things—escape being spheroidal. That's why asteroids and meteors are irregular.

The universe does allow loopholes, where objects warp themselves away from perfect roundness. Speedy rotation forces large fluffy planets like Saturn and Jupiter to bulge exotically at the equator. Even Earth has a polar diameter 27 miles shorter than a tunnel bored through its tropical dimension. Some stars, as well, spin so quickly they look as though squeezed in a giant vise. But such exceptions aside, it's a universe of spheres.

As to the planetary paths that caused so much self-inflicted grief to the Church, a circular orbit is really nothing more than a particular kind of el-

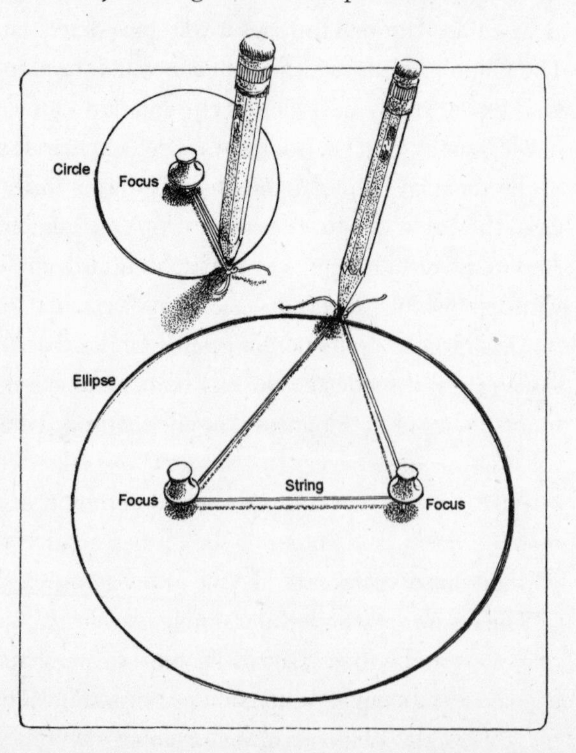

Upper left: *drawing a circle*. Lower right: *drawing an ellipse.*

lipse; it's really ellipses that deserve our attention when we explore the eternal tracks of the planets.

Unfortunately, ellipses seem intimidating to most people. Perhaps ovals come across as inelegant and hard to understand. Nevertheless, almost everything in the universe (we included) rushes through the cosmos along an elliptical avenue.

The best way to understand an ellipse is by drawing one. Hang in there; you'll come to love this. First we start with a circle. Drawing a circle entails pushing a thumbtack into a page and attaching it to a pencil with string; the rest is no challenge for anyone over the age of thirty-six months.

With an ellipse we need *two* thumbtacks; each is called a focus of the ellipse. Push them partway in, encircle both with a looped piece of string, insert the pencil, pull it tightly, and—bingo. An ellipse. See, not so hard. Move the thumbtacks farther apart and the ellipse you draw becomes more elliptical (or, to use the right term, *eccentric*). Bring the tacks closer together and it gets more circular. When the foci coincide, it *is* a circle.

Four centuries ago, Kepler figured out, after a decade of pondering the problem, that all planets travel in elliptical orbits, with *the sun occupying one focus.*

The other focus is just a vacant spot in space. This fact often bothers people: Nobody wants that other focus to be a mere mathematical point. It seems *something* should be there. Well, all right: When the era of interplanetary space travel arrives, we can construct a resort or restaurant there and call it The Focal Point. Catchy but unnecessary, for in real life the sun's occupying one focus is all that's necessary for a stable orbit. Every planet shares that solar focus, while each planet's second focus is somewhere else in space. So, how many *different* foci do the nine planets have? A good trivia question—try it on any amateur astronomer you know. It'll throw him for a, er, loop. (Answer: ten. Nine plus the shared one, the sun.)

But let's not lose our own focus. Our question is: Why shouldn't planetary orbits be circles? The better question is: Why should they? A perfectly round orbit changes into an elliptical one the moment the planet gains or loses the slightest amount of speed, which can happen because of collisions or gravitation perturbations by other bodies. In short, a circular

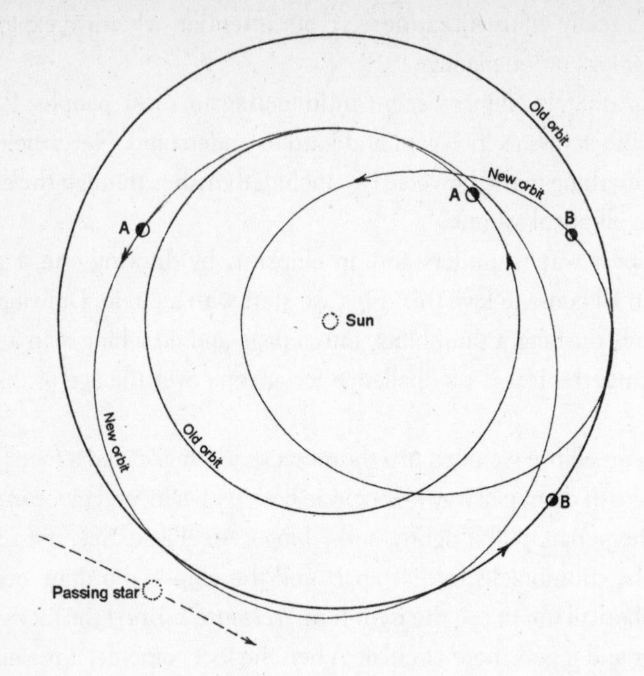

Imagine two planets, A and B, in circular orbits around their sun. A star passing near them at this moment will cause planet A to speed up and move farther from its sun; planet B will slow down and therefore fall toward its sun. As B nears the sun, it will speed up again, whereas A will slow down as it swings farther away. Both establish new elliptical orbits. The more massive the planet, the less its orbit will be perturbed.

orbit is just a specific kind of ellipse that is extremely unlikely to be maintained.

But if you're old-fashioned (like heavy-duty old-fashioned, with values from the sixteenth century) and abhor noncircular motion, your favorite planet must then be Venus, whose orbit is the roundest of all. Or, for the best circles anywhere in the solar system, you could stare at the paths traced by the innermost Galilean satellites of Jupiter—Io and Europa. Add Neptune's giant moon Triton and Saturn's Tethys and you've got the celestial Knights of the Round Table.

Going the other way, the most eccentric planetary orbit belongs to Pluto, whose distance from the sun varies with a 5-to-3 ratio. Throw in Mercury and Mars and you have the only planets whose paths would *look* oval to an alien staring down at the solar system from above.

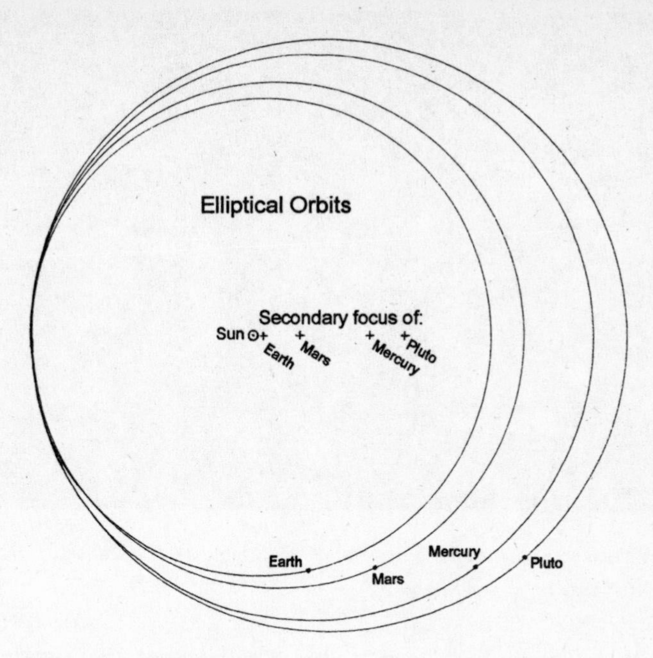

To compare the eccentricity of planetary orbits, we've superimposed the orbits of Pluto, Mercury, and Mars over Earth's orbit. Pluto's secondary focus is actually out near the orbit of Uranus!

Kepler also showed us that planets speed up when closest to the sun, and move slowest when farthest away. Therefore a planet like Venus or Neptune, with a very nearly circular orbit, cruises through space at a uniform speed. A planet like Mars, whose path is noticeably elliptical, ceaselessly speeds up and slows down like someone learning to drive.

Our own world has an orbit oval enough to carry us 3 million miles nearer the sun in January than July. In the lethargic days of early summer, we therefore travel a few thousand miles an hour slower. Since our clocks steadily mark the *average* position of the sun, while the real sun appears to vary its speed against the background stars because we vary ours, the consequent shifting of the sun relative to our clocks helps produce the changing times of sunrise and sunset. (Our tilted axis is the other cause of the yearly waltz of light and darkness.) The point is that Earth's lack of orbital circularity is no mere academic abstraction; it affects everyone.

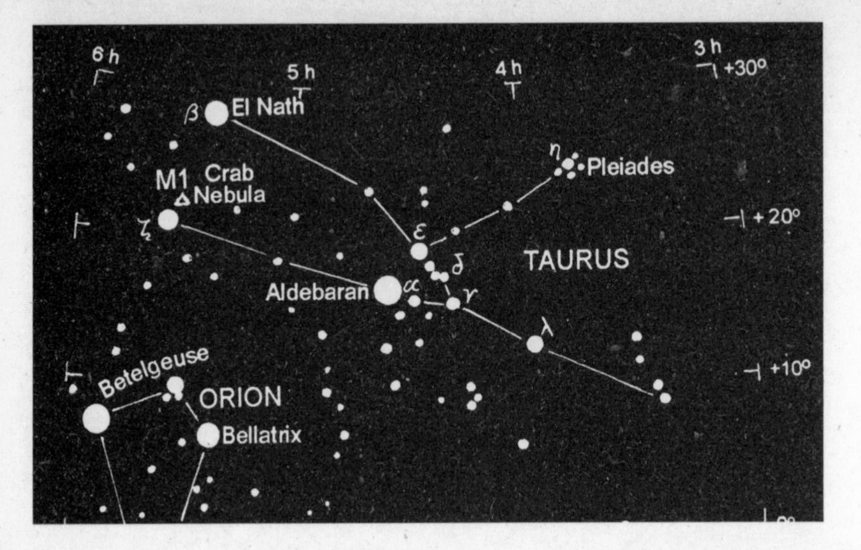

The Crab Nebula, passing high overhead in fall and winter, yields detail only in powerful telescopes. It blew up on the Fourth of July.

When describing the *shapes* of objects rather than their motions, we find we truly belong to a cosmos built of great balls of fire. From raindrops to the universe itself, reality is essentially spherical. Since our mandate is to explore the unusual, we must ignore most of the trillion blinding star-globes that sprinkle our galaxy like fiery ball bearings, and select a favorite handful of the most astounding. We'll start with the smallest—a 10-mile-diameter ball buried in the constellation Taurus, halfway up the eastern sky in early fall.

This jewel of the autumn first came to our planet's attention on the Fourth of July in the year 1054. That night, a brilliant starlike beacon appeared in the heavens and cast shadows for weeks. It was so dazzling, it remained visible even in broad daylight. Seen around the world and persisting for over a year, its position was chronicled in Chinese records. When the telescope age emerged and instruments were pointed at that spot of sky near the star Zeta Tauri, a chaotic nebula was seen, looking like the arms of a crab.

This Crab Nebula, as it came to be called after the telescope's invention, is the twisted remnant of a **supernova,** a massive exploded star whose debris continues to fly outward at 600 miles per second. The power needed to hurl the material of a million planet Earths at such speed is hard to imagine. No easier to grasp is the collapsed core of the dead star still sitting near its center.

Imploded into a tiny ball the size of Los Angeles, and hard as a diamond, it spins wildly at thirty-three revolutions each *second*. Because its magnetic axis points in our direction, and because subatomic particles are so accelerated by this wild spin they emit energy channeled along its magnetic poles, we receive a flash of light with each rotation.

This odd creature, flickering like a firefly out of control, is a pulsar. Members of the established zoo of celestial oddities today, pulsars were

The heart of a pulsar is a rapidly spinning neutron star. Its powerful magnetic field prevents the accretion disk from falling closer except at the magnetic poles. There the charged particles follow the lines of force to the star's surface, emitting X rays as they go.

first detected in the 1960's when faster films permitted shorter photographic exposures. Before that, the lengthy exposures necessary to capture these tiny crushed star remnants were much longer than their period of pulsation, which is why their astonishing variability went undetected for so long.

Each spoonful of the Crab pulsar weighs a *thousand tons.* A baseball of its material would outweigh a skyscraper. Space itself, and time, is warped by the Crab pulsar's extraordinary density, so that your wristwatch would run at a different rate when in its vicinity. But being punctual wouldn't be your first concern there. Its lighthouse flashes seem designed to warn away potential callers like an interstellar buoy. For anyone approaching too closely would be violently yanked downward by the awesome gravity and transformed into a thin film of jelly that would vaporize while spreading itself evenly around the pulsar's solid surface.

But oh, that spin, that dizzying spin. Its thirty-third-of-a-second rotations emit clockwork beeps of light and energy, a celestial beacon that has been slowing by thirty-seven billionths of a second per day.

Yet the Crab pulsar is not even the rotational record holder. Of the four hundred pulsars presently cataloged, about a dozen spin between 100 and 880 times per second, causing stars in their sky to appear not as moving dots but as perfectly circular, solid lines!

If a dazzling, spherical diamond whirling faster than the eye can see doesn't qualify as the strangest globe, be patient. For contrast, we can jump from one of the tiniest spheres to one of the largest.

It's the Cygnus Superbubble, discovered in the 1980's. Straight overhead at dusk in early fall, it's a jumbo version of bubbles commonly seen encircling stars that have suffered explosions in their outer layers, producing expanding shells of thin gas. The resulting so-called **planetary nebulas** (which have nothing to do with planets, except that early observers thought they looked like the newly found planets Uranus and Neptune) include such well-observed halos as the Ring Nebula seen through hobbyist telescopes on summer nights near the brilliant star Vega. Typically one-half light-year across, such huge, colorful spheres glow eerily in **forbidden radiation.**

The light is "forbidden" because, when analyzed spectroscopically, it displays the signature of something that we could not initially fathom, since by all rights it should not exist. The spooky greenish ring is a must-see through a telescope; one beholds a kind of light that cannot be duplicated on Earth, even in our laboratories. The glow comes from oxygen that has been stripped of two of its outer electrons, a state that oxygen normally avoids at all costs. What's needed to produce it is the very unearthly environment of a near vacuum, together with wide open spaces and high energies.

But even these strange planetary nebulas are amateur acts compared to the Cygnus Superbubble. Possibly caused by chain-reaction supernovas, this interstellar "shock wave" feeds its own existence by giving rise to new stars, which then explode, and on it goes in an expanding, self-perpetuating globe.

In this cutaway view of a large cloud of gas and dust, we see a chain reaction of supernovas in progress. The oldest supernovas (1) expelled a shell that collided with the surrounding cloud, spawning new stars, some of them massive enough to explode in their turn (2). These created an expanded sphere of massive stars that continue the process (3). A superbubble is in the making.

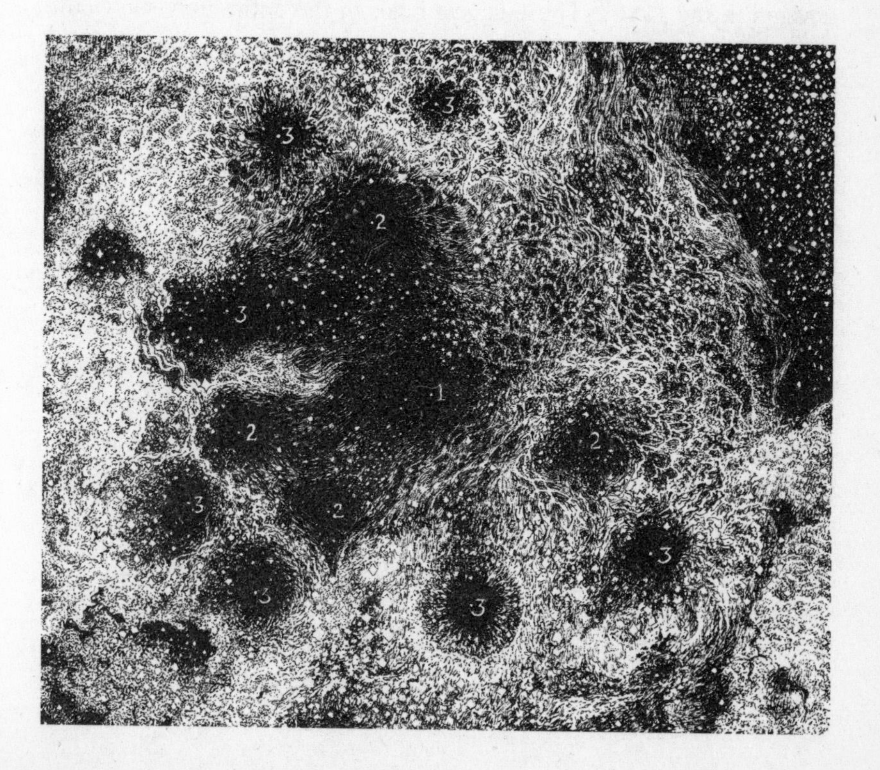

The Cygnus Superbubble has a diameter of 1,000 light-years, a dimension that cannot be treated casually. To appreciate such enormity, let's say we drop suns inside of it at the rate of a billion a second. With the sun itself a million times larger than our planet Earth by volume, filling anything with suns at such an impressive billion-per-second rate has to make a serious dent at plugging any cavity. Yet the project would take 9 trillion years—some five hundred times longer than the universe has existed.

Just traversing it would be a challenge. If our rocket technology ever improves to the point where we can travel a thousand times faster than we can today, the Pan-Bubble route would still take 6 million years. You'd run out of things to read.

In 1993, even larger bubbles were found in the distant galaxy NGC 4631, probably also caused by supernovas. The largest spans some 10,000 light-years, so that it could swallow up even the Cygnus Superbubble a thousand times over.

But perhaps we're getting jaded with mere unimaginable enormity; we're ready for another giant leap into the fantastic bubble kingdom uncovered in the 1980's. These largest holes in the entire universe would overshadow the Cygnus Superbubble the way a ball 8,000 feet in diameter compares to a marble. They are the largest structures ever detected.

Large bubbles of emptiness shouldn't amaze us; the cosmos, after all, is closer to "nothing" than our best laboratory vacuums. Smear everything out and there are fewer than one hundred atoms per gallon of space: nothing. The richness of our neighborhood, the creamy fullness of whales, skyscrapers, and even clouds, is an exception, an oasis in the great enchanting Emptiness that is our home.

But the newly discovered intergalactic voids are impressive even in a cosmos accustomed to vacuity. Unimaginable empty spheres defined by curving sheets of galaxy clusters, they span 200 million light-years across, and more. The universe, then, for reasons no one understands, is shaped like a sponge on the largest scale. The smaller spheres of stars, planets, and globular galaxies encircle vast balls of very nearly nothing at all.

It's enough to make one's head go around in circles. And it also provides a giant headache for the Big Bang theory, since not enough time has elapsed since the universe's presumed genesis for such grand formations to develop.

Galaxies cluster around great empty cavities, as if the universe were built like a sponge.

How can we top this? This ballgame of one-upmanship can go only one place: to the universe itself.

The formula for the volume of a sphere is $\frac{4}{3}\pi r^3$, where r is the radius. You may hate math, but you cannot ignore how very easy it is to calculate how big is the interior of any ball, even the entire cosmos. This exercise is necessary if we're to make the volume of Everything meaningful.

Years ago, stores would sometimes display giant jars filled with marbles or jellybeans. You'd win a prize if you correctly guessed how many they contained. Perhaps that's why so many of us lose sleep with this nagging question: If the entire universe were hollow, how many marbles would it take to fill it? It's one of those issues that just won't go away. And a question with more relevance than you might think, since the size of a marble is exactly the dimension our planet Earth would assume if it collapsed just enough to achieve black hole status. So, how many marbles would be needed to fill the cosmos?

The answer: 224,900,000,000,000,000,000,000,000,000,000,000,000,000,000,000,000,000,000,000,000,000,000,000,000,000,000,000,000,000,000,000,000! That's assuming the universe ends abruptly at the farthest quasar we can presently detect, which is unlikely. So no marble factory, no matter how efficient, need worry about running out of storage space, even if the crew works overtime.

But if you were consumed with cramming the universe with marbles, and could somehow produce a trillion marbles every trillionth of a second, and you had a trillion factories doing all this simultaneously, and had this obsessive setup operating on every star of every galaxy, it would still take 7,131,500,000,000 years, hundreds of times longer than the universe has existed. And meanwhile, of course, you'd never be able to keep up with the universe's expansion.

When astronomers dismiss astrology and other pseudosciences, some people reply, "But anything's possible, right?" This marble business is a good illustration that some things are *not* possible. We can safely say it's impossible to manufacture more marbles than there is space for.

In any event, all these fantasies about empires of marbles have, in retrospect, probably not succeeded in the slightest in making the cosmos more graspable. Even if we reduced the figure by twenty-seven zeros by expressing the number of *planet Earths* that would fill the universe, the spheres that compose reality still lie far beyond our powers of visualization.

If we struggle too hard to picture it all, we could lose our marbles.

Harvest Moon

It's the same story every year. "Harvest Moon tonight!" says the TV weatherperson, and all across the country people think they know what it means.

Some assume it has no observational significance, that "Harvest Moon" is an antiquated label for the autumn full moon—just another archaic name like April's "Grass Moon" or March's "Sap Moon." It is merely, they reason, an obsolete term left over from simpler times when moonlight still had relevance to everyday life.

Others imagine just the opposite, that the Harvest Moon looks special in some way: bigger or redder or higher or—*something.*

Not even close!

No other moon is as famous or as misunderstood, a combination that qualifies the Harvest Moon as a full moon worthy of our investigation.

We can immediately dispense with the notion that it appears different from other full moons. It doesn't. The only way the moon ever noticeably changes its naked-eye appearance is by wandering closer to or farther from Earth in its elliptical orbit. Any almanac will list the monthly apogee (far point) of the moon, as well as the date of nearest approach, or perigee. The variation is noticeable, frequently amounting to a 12 percent change.

But neither extreme corresponds to the time of the Harvest Moon, except during the odd year when the laws of chance conspire.

Then there's our atmosphere with its own bag of tricks. In the haze and humidity of summer, the moon looks more orange, especially since the full moon then travels its lowest path of the year and spends more of its time in the thicker air near the horizon. It's even been suggested that the term *honeymoon* derives from the amber-colored full moons of June, the traditional month for weddings.

Conversely, full moons near the winter solstice are the year's highest, and usually shine through drier, crisper air, making them whiter and a bit brighter than most. But September's Harvest Moon? No, nothing special there.

Then you've got the famous **moon illusion.** Perhaps the most powerful mirage of all, it makes the moon seem enormous when it's down near the horizon, a visual effect caused by its proximity to foreground terrestrial objects. But again, it acts on all moons equally. This bit of celestial sleight-of-hand may come from our lifelong perception of objects near the horizon as being distant and hence larger, which grants them a bigger psychological size than when the moon is high overhead and dwarfed by the sky's vastness.

So if it doesn't *look* special, what's the fuss?

Perspective makes objects appear larger if they are near the horizon than anywhere else. The circles representing the moon and the front wheel of the motorcycle are the same size.

The eastern sky in spring, showing the moon's position on successive evenings at sunset

The eastern sky in autumn, again showing the moon's position on successive evenings at sunset. The moon moves the same distance each day, but at a slant that keeps it closer to the horizon, making it rise at nearly the same time night after night.

It all boils down to that old mischief maker, the tilt of Earth. Our slanted spin makes the moon's path seem to wobble like a tire out of balance, which in turn affects its rising and setting. On average the moon comes up about fifty minutes later each night, but in spring the full moon's rising is delayed by its steeper angle to the horizon. Then, an hour and a half can pass between moonrises on successive nights, the exact interval depending on the observer's latitude. But there is no special name for this springtime phenomenon of the moon's rising one night, and then practically vanishing the next.

In the autumn it's the opposite. Each night the moon advances the same 13 degrees in its orbit, but at a skinny slant to the eastern horizon, placing it barely below the skyline so that it rises just twenty-five minutes later.

That's it. That's the whole deal. The Harvest Moon doesn't look different at all; it's simply a series of evenings when the full moon rises most nearly at the same time.

A view of Earth from space near the time of the autumnal equinox. If we align the moon's orbit with 42 degrees south, as at left, we see that the moon moves almost perpendicularly to the horizon. At right we align the moon's orbit with 42 degrees north and see that it crosses the horizon at a shallow angle. In spring the angles are reversed. These changing angles produce the Harvest Moon effect.

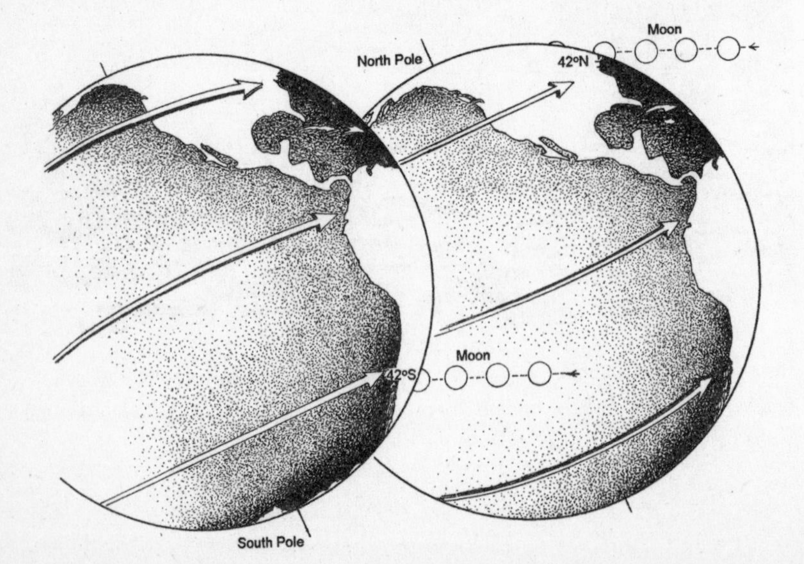

The result? Picture it: The harried farmer trying to finish harvesting chores is running out of daylight. The sun sets and—bingo—just then the full moon rises to add welcome light. And this goes on for several nights in a row.

So the Harvest Moon is more of an *event* than a *thing*. It's the phenomenon of the full (and nearly full) moon rising at around sunset for several consecutive evenings. The following table lists the dates of the Harvest Moon, but to observe the Harvest Moon *effect* you must watch the moonrises for one or two nights before and afterward.

Harvest Moons

P indicates years in which the Harvest Moon falls within a day of lunar perigee, yielding an exceptionally large Harvest Moon.

1995	Sept. 9
1996	Sept. 27
1997	Sept. 16 *P*
1998	Oct. 5 *P*
1999	Sept. 25
2000	Sept. 13
2001	Oct. 2
2002	Sept. 21
2003	Sept. 10
2004	Sept. 28
2005	Sept. 18
2006	Oct. 7
2007	Sept. 26
2008	Sept. 15

Traditionally deemed the full moon nearest to the autumn equinox, when days and nights are roughly equal, the Harvest Moon can occur as early as September 8 or as late as October 7, and either of these extremes causes the effect to become somewhat diluted. Those Harvest Moons occurring closest to the September equinox (1999 and 2002) have the additional benefit of rising almost precisely due east and setting due west, a nice symmetry. At those times due west blazes with the fire of the setting sun just as the precise eastern point is marked by the rising of the full moon. Moreover, this dramatic equilibrium is observed everywhere in the world.

But if you're not too fussy, *every* Harvest Moon comes up very nearly due east at around the same time the sun sets at nearly due west; certainly the cardinal directions are being indicated more accurately and strikingly than by reading a compass. It's the sort of effect primitive people used to beat drums about. While few in today's cities seem inclined to make fires or perform rituals, we can at least glance at the equinoctial sunset and Harvest Moonrise and picture the centuries of ceremony when such lunar and solar positions were marked by idols or stones, à la Stonehenge.

We can also, perhaps, take in the universality of the experience. Most other celestial events are better seen at one location or another. Some—solar eclipses, for example—are visible only from specific geographical regions. But the Harvest Moon, occurring as it does at around the time of the equinox, floats over the equator as does the sun, which ensures that both are visible everywhere. The phenomenon is worldwide.

Incidentally, that "day and night equal everywhere" business, repeated as gospel at every equinox, may have a nice ring to it, but there is no date on which such a phenomenon occurs worldwide. The spoiler is atmospheric refraction, which bends the horizon sun's image upward by half a degree, bestowing a few extra complimentary minutes of sunshine. The equinox therefore favors sunlight over darkness. True equality of day and night happens nearly a week later in the Northern Hemisphere, and almost a week before the fall equinox for those living south of the equator. But it's close enough to provide an excuse for pagan ceremony; the equinox represents approximate universal equality to within twenty minutes or so.

And the sky's harmony resonates in other ways as well. South of the equator the Harvest Moon rises up and to the left, while the rest of humanity sees it come up and to the right, producing an almost poetic planetwide symmetry. It's as if Heaven and Earth paused for one ephemeral breath, balanced and motionless, before rushing headlong toward the northern winter.

Northern Exposure

In most people's minds, the aurora borealis is linked with an arctic image—a sight for marginally sane scientists working in snowdrifts surrounded by polar bears. But the riotous curtains of color are not confined to distant frozen lands. To see the eerie spectacle you need only journey to the nearest dark sky.

That's because the displays are part of an immense luminous ring centered over Earth's *magnetic* poles, where compasses point, rather than those *geographic* poles to which explorers crusade. The closer you are to this **auroral oval,** the brighter and more frequent your northern lights will be. By sheer good fortune for those in the United States, the magnetic North Pole lies in nearby Canada, due north of the Texas Panhandle at latitude 77—much closer to us than the pole of rotation. People at our latitude in Europe or Asia are some 1,800 miles farther from the auroral oval, and it makes all the difference.

Auroras occur monthly or more often over New York and the redwood forests of California. European sites equally far north are out of luck: Rome, whose latitude matches New York's, sees them only once every three to five years. Nor is the City of Light a good place for northern lights; they're as infrequent in Paris as in Florida (1 percent of all nights),

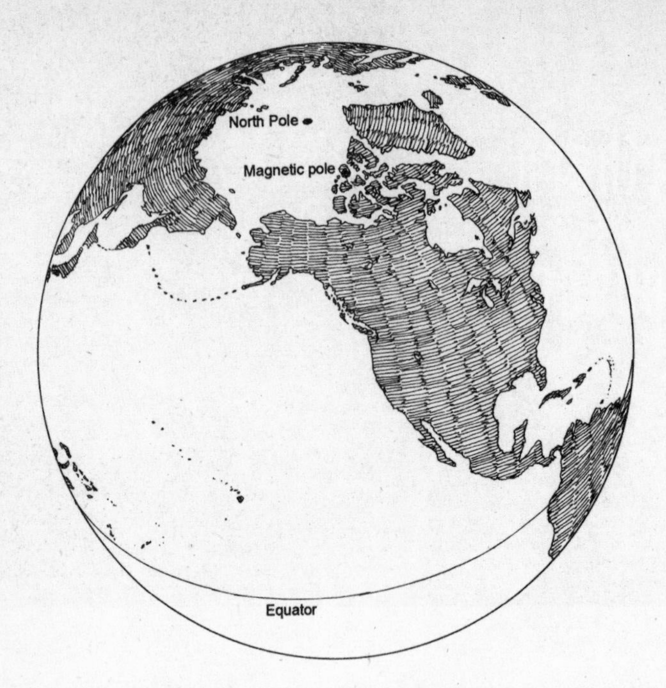

The magnetic pole is presently more than 800 miles south of the rotational North Pole, in the islands of northernmost Canada.

even though the Eiffel Tower sits at the same latitude as Vancouver, where they blaze every week or so.

None of this matters if you live under the milky mess that paints today's urban ceilings. But in rural areas, especially in the northern third of the United States, majestic displays frequently adorn backyard skies. One North Dakota amateur recorded eight hundred in a decade. If your site's good enough to show the Milky Way, you'll see auroras as well.

So forget dog sleds and igloos. What's needed is the crisp conditions often found in the autumn, and nights when the moon is skinny or absent. Auroras are a bit more common in the fall for another reason as well: It's easier for a display to appear when our planet is tipped sideways to the sun, around the equinoxes. A little advance notice might be nice, too—possibly gained if you have a telescope with a solar filter, and see a giant sunspot group carried toward the center of the sun's disk by its twenty-seven-day rotation.

When the storm faces us it may send a blast of charged particles across Earth's magnetic field, generating electricity that excites atmospheric

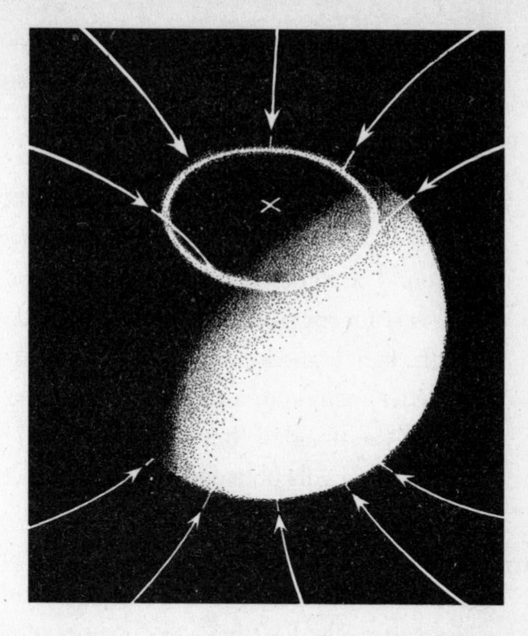

An image of Earth in ultraviolet light shows a ring of auroras surrounding the magnetic pole (marked with a cross). The aurora forms in this ring because the charged particles from the solar wind that induce it are compelled to follow lines of magnetic force (arrows).

atoms 100 miles overhead. The high-voltage jolt zaps their electrons into more energetic states. As the electrons cascade back down to their preferred, lower orbits, they quietly emit sprinkles of green or red light.

This is major-league energy, dwarfing the power of lightning: 20 million amps at around 50,000 volts is typically channeled into the auroral oval, lighting it up like the gases in streetlamps.

But the only question that really matters is: When will this happen? Nobody needs to be sold on this dramatic if eerie Off Broadway show; people simply want to know when and where to look. Alas, the northern lights usually ignore earthly calendars and march to a different, more celestial drummer; they follow the violent, internal vagaries of the sun itself.

The sun routinely unleashes a wind of charged subatomic particles that, traveling a million miles an hour, takes four days to reach us. But this normal blast of material is deflected by our protective magnetic shield, and does not produce celestial fireworks. What's needed is an extra-intense burst, the kind that sunspots (solar storms) are particularly capable of providing. Their hurricane gusts of broken bits of atoms tease our magnetic shield into generating fierce voltages. That's why the ebb and flow of the eleven-year sunspot cycle correlates well with displays of the northern lights.

(Actually it's a twenty-two-year cycle if you pay attention to the magnetic orientation of the sunspots; in every other cycle the positive and negative charges of the spots reverse themselves, with consequences that are anybody's guess.)

The mid-nineties found the sun in a quiet period. The closing years of the century and opening moments of the new one should give rise to a fresh round of numerous auroral fireworks, as the cycle heats up anew.

It is easy for any small telescope, equipped with a proper filter that cuts down the light *before it enters the instrument,* to keep track of solar storms. Even better, many aurora-causing sunspots are so large that they stand out clearly to the unaided (but filter-protected) eye; they dwarf our planet Earth. A safe, easily obtained filter with which to view them is a lens for number 14 welders' goggles, available from any welding supply store. Just a two-dollar replacement filter will do; you don't need the whole mask unless you'll be repairing your car at the same time.

But you really don't have to do your own solar observing unless you want to. Current details on the sun's shenanigans can be had by dialing (303) 497-3235. Imagine: The sun, by phone. It's an inexpensive call if dialed by night—as if the rate were deliberately reduced because the object of interest is absent. As a bonus, you'll also get information about the condition of Earth's magnetic field. If you're the type who worries about things, this will provide one more aspect of life of which you can keep track: your planet's magnetism.

Now that you know how to get advance notice of an aurora, here's another tip: Forget it. Predicted auroras seldom materialize, while the best displays often arrive unheralded. The optimum method, if you live in rural or darker suburban settings, is simply to get in the habit of glancing at the sky while returning home each night. Establish in advance the normal background glow to the north of your house. That way an anomalously bright sky will alert you immediately, not just to auroras, but to outdoor parties at your neighbor's.

Your local astronomy club could organize an **aurora hotline** similar to the one in my town; if someone spots the alien gleam, others are phoned. After years of experience with such a telephone chain, my advice to those establishing one is: Be sure members have the approval of everyone in their household. After spotting a beautiful midnight aurora, the last thing one wants is to go indoors and be tied to the phone, only to receive an insult from a sleepy, grumpy voice.

The most common type of aurora is also the variety that often escapes notice—a pale greenish or ruby patch splaying across the northern sky. But in reality, no two auroras are identical. They can exhibit blotches, rays, bands, arcs, or eerie curtains that slowly undulate like rustling draperies. They can be nearly motionless, pulsate leisurely, or flicker rapidly with five new scenes every second. Their silence is unearthly, and so are the bizarre hisses or crackles that have occasionally been reported to accompany major arrays. Such auroral noises are controversial; scientists who accept their existence suggest that some people may be able to sense the huge electrical charges that course along the ground beneath bright displays.

Even a minor aurora can quickly evolve into one of the more active forms, like the dazzling light show on March 13, 1989, that embraced the whole sky, and was seen from nearly the entire United States. While that performance occurred during a peak period of sunspot activity when auroras normally intensify, these alchemists of surprise also develop in times of solar calm, such as in the mid-nineties. They can happen any night.

The intense solar violence responsible for an aurora affects us in other ways as well. The celestial fury that created the

The aurora's ghostly glow can assume many forms. No two displays are identical.

spectacular 1989 aurora caused such fluctuations in Earth's magnetic field that voltage surges coursed through power lines. Transformers blew out, plunging millions of people into a total blackout in Montreal and much of Ontario. Garage doors throughout North America spontaneously opened and closed for hours!

Earth's upper atmosphere heats and expands at such a time, forcing satellites to glide through increased resistance, slowing them down, dropping them to lower orbits, and even causing some to require constant commands from Earth to keep from tumbling out of control. And yet sunspots are necessary. The absence of dark spots on the sun's face paradoxically reduces the energy sent toward Earth, making us colder.

There is much evidence that solar cycles can disintegrate, with serious consequences. A lifetime-long absence of sunspots plagued our planet from about 1640 until 1710. Now called the **Maunder Minimum,** this total loss of the sunspot cycle was a period of extraordinary cold and hardship. The English Channel froze solid, glaciers started heading southward, and there was widespread suffering. Nobody at the time linked the icy misery with the sun's appearance, or even cared that the sun was virtually free of spots. Some astronomers even started to question whether extensive sunspots had ever existed, as those earlier

The records of sunspot activity over the last 350 years show an extraordinary lull in the second half of the seventeenth century, called the Maunder Minimum. This coincided with extremely cold weather over the same period.

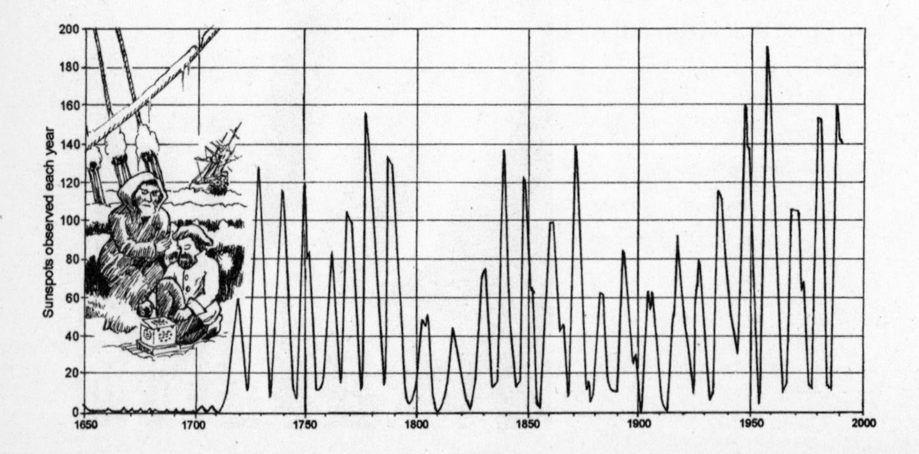

observations from the first years after the telescope's invention faded further into history. Nor, with the passage of decades, did people seem to wonder why those grand lights of the north that had been so beautiful in their childhood had vanished.

Auroras, then, present us with more than an awesome light show. They provide a kind of reassurance that all is still right with the sun, that the enigmatic disruptions of the sun's rhythms have fallen idle—hopefully for many centuries to come.

Leaving Home

A space probe quietly crosses the night sky: technology wrought by a million minds distilled into a drifting dot of light.

No matter that people have walked the moon and robots have explored seven planets. It remains exhilarating, even implausible, seeming like some fantastic dream, to be alive in this unparalleled era of cosmic exploration that began one autumn day in 1957.

Of course, many are greedy for more—much more. The Planetary Society's one hundred thousand members are a lobbying group pushing for

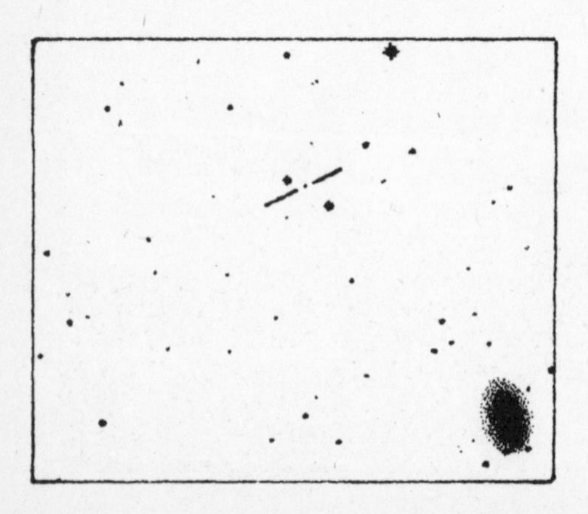

Quasar 1228 + 42W1 (between the two lines). Although it is a galaxy, it is so distant that we can see only its bright core, which appears starlike. Creatures there cannot see us at all, no matter how powerful their telescopes. The image of our birth will not reach them for another 5 billion years.

full-steam-ahead voyages into ever deeper space. Their opponents insist those resources be allocated toward earthly needs. To evaluate such conflicting goals, we need some realistic understanding of that most astonishing fact of the twentieth century: space travel.

Most people are oblivious to the ABC's of space science. They see televised scenes of astronauts weightlessly floating in orbit and believe they've escaped Earth's gravity. They think being in space is great fun, and not too unhealthful. They expect colonies on the moon within the next century.

No. No. And (probably) no. So, let's follow the most common misconceptions right from the launching pad. . . .

We visualize rockets launched into space straight up or nearly so. But the initial fiery rise is vertical for less than the time it took to read this page. The idea is to get through the thick lower atmosphere as quickly as possible and then tilt to the east. When the rocket reaches its desired height it travels perfectly sideways!

This wonderfully choreographed tilt is obvious in televised launches, when the craft is high overhead, spewing flame. People assume the on-its-

Ten seconds after lift-off, the space shuttle is already beginning to tilt out of the vertical. It will roll about its long axis and pitch onto its belly for horizontal flight.

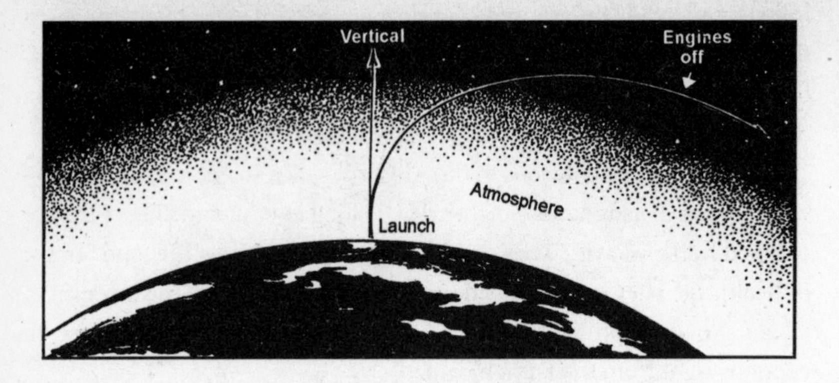

An orbiter's course after lift-off quickly tilts eastward. With engines off, the craft is essentially falling, but Earth's surface keeps curving away from its fall.

side orientation of the distant, departing rocket is simply some artifact of the camera angle. Not so; it's *really* going sideways. When it reaches orbital height, the engines are shut off and the rocket is simply allowed to fall. But its horizontal speed of five or six miles per second combined with the vertical pull of gravity causes the ship to plummet in an arc, just as a baseball thrown sideways falls in a curving path.

The arc of the craft's freefall just matches the curve of the planet below, and so the plunge continues without end. That's what being in orbit means. And just as you would feel weightless in an elevator whose cable had snapped, the astronauts feel weightless for one simple reason: They're falling!

Gravity? There's plenty of gravity up there. If we built a skyscraper as tall as the astronauts orbit (200 to 250 miles) and stood on the roof, we'd still weigh about the same as we did on the ground. A couple of hundred miles, after all, is not very far above a planet 8,000 miles in diameter.

As for the fun of feeling weightless—half of all astronauts become extremely nauseated and remain so for days. The human body uses gravity as an integral part of the circulatory system. Gravity's absence causes physical discomfort, and its effects are none too flattering besides. Narcissists should stay home: Blood and fluids accumulate in the face, bestowing a puffy, ghoulish look. Over the long term, calcium loss inevitably occurs in the bones, and some of this is irreversible.

Shuttle's height to scale

Orbiting astronauts are much nearer to home than popularly thought.

Such physical liabilities would be somewhat alleviated on the moon or other celestial bodies where gravity was present. But that's hardly enough incentive to press for extraterrestrial colonization. The *Apollo* finding that the moon is bone-dry means that every drop of water, so vital to human life, must be hauled there. The utter lack of air forces us to bring cumbersome oxygen production equipment. The boiling 260-degree-Fahrenheit daytime temperatures require powerful and energy-exorbitant air conditioning. The night's frozen, minus-250-degree readings dictate prodigious heating requirements during the two-week-long lunar night that cannot be met with energy from the vanished sun. No hydro power is available, nor is there oil or coal, and in any event no energy would be produced using a combustion process because it would consume precious oxygen. That leaves nuclear: A moon base would be energized with atomic power, bringing its own set of problems.

The moon's absence of essentials, from raw materials to pharmaceuticals, would require constant resupply from Earth. Everything from batteries to water to cheese would cost at least ten thousand dollars a pound for transport alone. Want a little ice in your ginger ale? That'll be an extra five thousand dollars, please.

What could possibly justify such an expensive colony? No gold has been found on the moon, nor diamonds nor uranium. The possible value of weightless or airless manufacturing could be accomplished as readily in Earth orbit, a thousand times closer to home. A scientific or astronomical station could be easily run with robotic, automated equipment.

What reason, then, for a lunar colony, and who would pay for it? The answers are none, and nobody. That's why you won't see one in your lifetime, and most likely neither will your grandchildren.

Mars, however, is another story. While still not exactly a vacation in Tahiti, a Martian stay would at least offer water and oxygen. Or, more accurately, ice and oxygen-bearing minerals. Mars' temperatures (55 degrees during a perfect summer day, and a rather nippy minus 180 at night) are far less extreme than those on the moon. With the right technology, an enclosed colony probably could remain self-sufficient in many fundamental life requirements. (But much would be lacking, like a summer breeze or the sound of birds singing at dawn. . . .) A Mars colony would

The view across Mars' Noctis Labyrinthus. The high plateau of Mars is split by numerous faults, which here have opened up into a network of deep canyons with sheer walls subject to landslides. The volcano Mount Pavonis looms on the horizon, 300 miles distant; its peak soars to a height of 13 miles, causing clouds to condense on its slopes.

remain even less hospitable than a station at the polar regions of our own planet, where people aren't exactly lining up to live.

At least in the arctic, one can breathe the air and find water without a hassle. And yet, who can say how advances in technology may change the picture? Limited human exploration of both the moon and Mars will surely occur in the twenty-first century. Permanent colonies, however, are extremely unlikely in the lifetime of anyone now living, and perhaps—as heretical as it sounds—might never make sense. After all, there are many earthly settings that we could inhabit, given lots of funding and motive, and choose not to, such as the bottoms of oceans and the summits of Himalayan peaks. Both are easier to survive in and less expensive to maintain than Martian or lunar bases, and yet we don't live there because there just doesn't seem any point in doing so. Given that extraplanetary colonization is such a staple of science fiction and our collective future fantasies, the suggestion that it may not prove feasible is never even whispered.

So far we've considered only the friendliest places in the solar system. Things go downhill when we look elsewhere. The moon is a honeymoon compared to the nearest planet, Venus. There the temperature simmers at an admirably uniform 850 degrees, with a pressure forty-five times greater than the inside of a pressure cooker. The air is unbreathable carbon dioxide, liberally sprinkled with concentrated sulfuric acid droplets: a drizzle from hell.

The mind of a fiend would be hard pressed to conjure a torture chamber as horrible as our "sister planet." Certainly Venus will never allow human visitation, let alone colonization. Not in a hundred years, not in a thousand.

Mercury is almost as inhospitable, although its polar region, replete with possible ice discovered in 1991, could offer a more temperate site, but with a notable lack of such amenities as an atmosphere of any kind.

That leaves us to contemplate the Jovian planets, the ones beyond Mars. Alas, they all have in common this major flaw: no surface! An astronaut attempting to land would find the ship traveling through a gaseous brew of hydrogen and ammonia before arriving at a slushy liquidy level that would simply swallow the craft. The crew, during its brief period of consciousness, would be sparkling clean from the ammonia, but that would be the only positive entry in the ship's log.

Like tourists with time and money but facing a boarded-up travel agency, we find ourselves with no place to go (except possibly to Mars). Of course the universe is vastly larger than our own solar system of familiar planets. But interstellar space presents a barrier that is nothing to sneeze at: 26 trillion miles of nothing, to be exact. That's the distance to the nearest possible planets outside the solar system: seven thousand times farther away than Pluto. At our fastest current speed, using a gravitational slingshot boost by carefully whipping the craft past the giant planets Jupiter and Saturn, we could reach those triple suns of the Alpha Centauri system after a journey of twenty thousand years. However, we have absolutely no evidence of any planets in that region. Twenty millennia is a long trip to take to discover there's nothing to photograph when you get there.

The nearest sun beyond our own that does show indirect evidence of a planetary system, Barnard's star (a binocular object in the southwestern sky in early fall), is 50 percent farther away—so better make that a thirty-thousand-year trip. Meaning a sixty-thousand-year round trip. You couldn't count on a welcoming party when you returned.

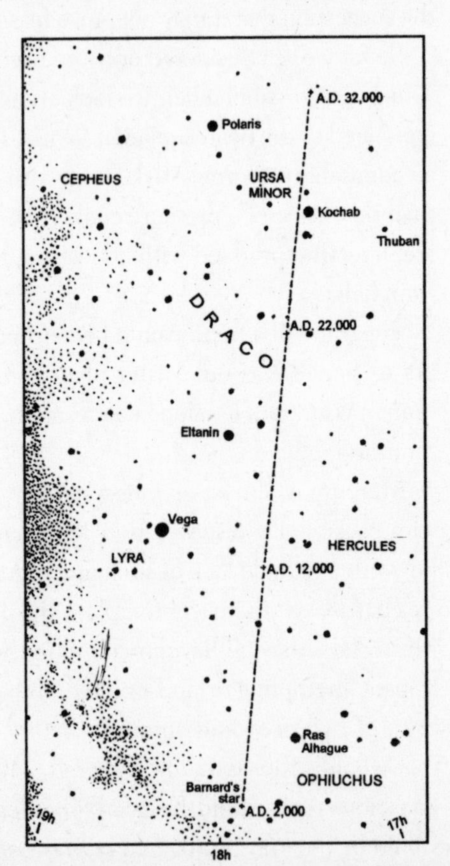

Barnard's star, just six light-years away, is a moving target. If we headed its way today at current speeds, we'd have to aim toward Polaris to intercept it. Moving north 1 degree every 280 years, it will have migrated to a position near the North Star during our 30,000-year voyage.

Interstellar travel is just not a viable idea at the present time. Nobody's even seriously proposing it, even for an unmanned probe. It could be done; we have already achieved speeds capable of escaping from the solar system, and four spacecraft have permanently departed into interstellar space. But these, *Pioneers 10* and *11* and *Voyagers 1* and 2, are heading out in arbitrary directions, having finished their planetary missions. They have not been aimed toward a particular star, nor will they reach a star's vicinity any time soon. The speediest *Voyager* will "encounter" the dim red star Ross 248 in forty thousand years, but what kind of visit will that be? The spacecraft will, at its nearest, pass some 1.7 light-years (40 million Earth-moon separations) from the star. No conceivable technology, if advanced creatures lived there, could detect the faraway bus-sized probe. Plus, of course, it will be dark and dead, its power supply exhausted four hundred centuries in the distant past. Despite the *Voyagers'* romantic portage of disks depicting city scenes and Chuck Berry music (okay, not so romantic), they are not rockets to the stars.

It would make no sense to embark on such a mission now. A craft launched today toward Barnard's star would be quickly overtaken by a rocket with twice the speed, leaving Earth a century hence. Even a ship launched two thousand years in the future, but that managed a mere

Current speed (represented by Voyager) offers little incentive for star travel. We'll improve in time.

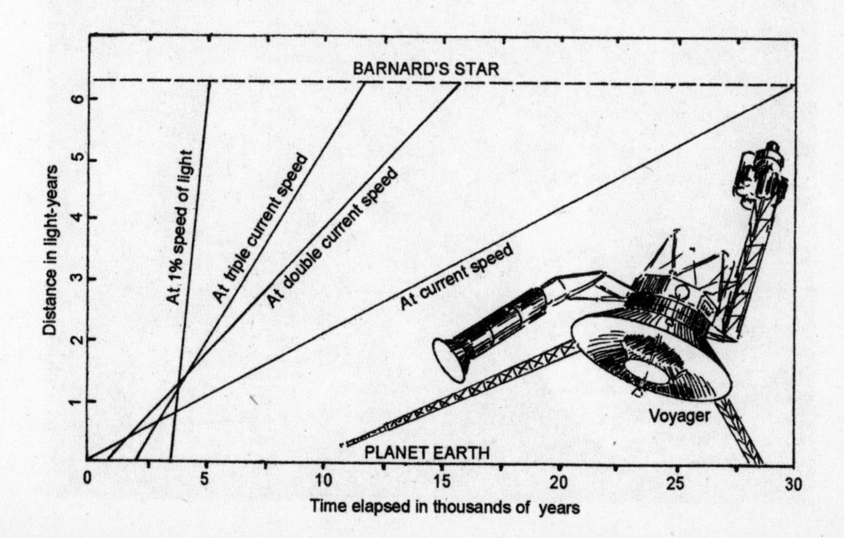

tripling of today's velocities, would arrive at the star eighteen thousand years sooner!

By this reasoning, it would never pay to set sail for the stars until we can achieve near-light-speed capability. Actual light speed is impossible because infinite energy would be needed; the ship would then weigh more than the entire universe. But there's no theoretical reason why a speed just a whisker below that of light shouldn't be attained, probably by technology not now dreamed of.

Such a velocity (186,282 miles per second) would further expedite human travel to the stars because of the time dilation factor of Einstein's relativity, which slows the travelers' aging process. At 99.999999999 percent of light speed, the crew would experience the passage of only a single year during a voyage to the center of the Galaxy. Simultaneously, 22,360 years and 8 months would elapse back on Earth.

Achieving virtual light speed in a manner physically tolerable to a human crew would require a careful acceleration. An ever-increasing speed that

Off to Alpha Centauri, our spacecraft has passed just outside of Jupiter to get a slingshot boost; we're now passing the moons Callisto (nearer) *and Ganymede. In the Milky Way above Callisto you can pick out the bright W of Cassiopeia. Look at it closely, and then refer to the next illustration.*

produced a 1-G force would be a good, logical pace because such a G pressure mimics the natural gravitational experience of Earth. One year of such an acceleration would achieve near-light velocity. Any attempt to hurry things would be physiologically unbearable for long periods, while a dash to light speed in less than a week would transform the crew into a film of gelatin. So much for science fiction—like the crew of the *Enterprise* reaching light speed in moments (and without bothering with seatbelts).

Slowing down when approaching the starry destination would take another year, but the round trip could easily be accomplished in a human lifetime. Energy requirements for reaching such fantastic speed (a velocity equal to thirty round trips between Boston and London in a single second) would be utterly impractical today. The fuel for such an enterprise would have to be far more potent than even nuclear power, and yet remain safe for the occupants. Even a hydrogen bomb, after all, converts just 0.7 percent of its fuel into energy—still somewhat inefficient for interstellar travel—while simultaneously emitting lethal doses of radiation.

A device that theoretically could do the job is an antimatter engine. Antimatter can be anything—water, sawdust, cigars—with this single departure from ordinary matter: Its electrical charge is reversed. Its atoms have nuclei that are negatively charged instead of positive, and the orbiting electrons are positive instead of negative; that is, they're positrons.

The same part of the Milky Way, now seen from Alpha Centauri. The W is slightly distorted, and above it we see a bright new star: our sun, 4.3 light-years away.

Another nearby sunlike star is Tau Ceti, 11 light-years distant. In its vicinity, we might find a view like this. Orion is still recognizable (top center), *though the ecliptic plane of this solar system is tilted about 70 degrees from ours. Sirius and Procyon no longer accompany Orion since we passed them on the way.*

We don't see any antimatter around us in nature, for good reason. When it contacts ordinary matter, it annihilates itself and the unfortunate object it encounters, in an explosive flash derived from the 100 percent conversion of mass to energy. This makes antimatter the most efficient possible producer of power. It has another attractive quality: It can be anything. Chewing gum wrappers, old tires, bounced checks, even dead bodies—anything's as good as anything else.

Unfortunately, half the energy produced is in the form of useless and dangerous neutrinos, which flash through the thickest steel as if it were fog. These nearly massless subatomic particles can be neither contained nor focused. Nothing can reflect or guide them, and no energy field influences their motion.

A more basic problem surrounding antimatter involves problems of storage. Even with its astonishing potency, several hundred tons of this fuel would be needed for interstellar travel. You can't keep it in any tank because it will annihilate the walls. But here, at least, an obvious if technologically challenging solution offers itself: magnetic confinement.

You'd somehow have to manufacture and then ionize the antimatter so that it's left with a net charge, rather than being electrically neutral. Then

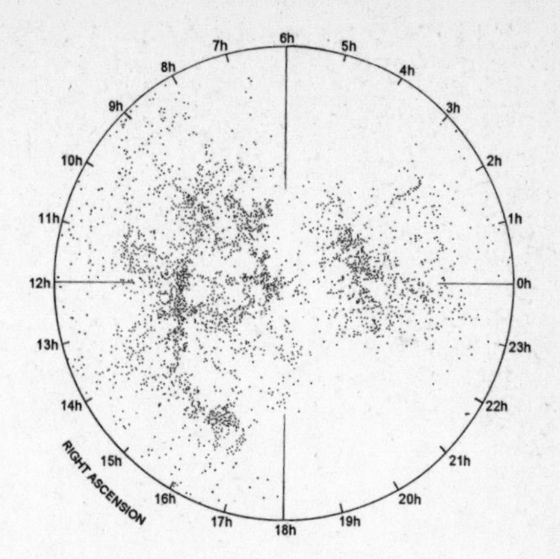

Galaxies within 600 million light-years of us (center). The two empty areas are where the Milky Way blocks our view. AFTER MARGARET J. GELLER AND JOHN P. HUCHRA OF THE HARVARD-SMITHSONIAN CENTER FOR ASTROPHYSICS

it would be susceptible to containment by magnetism. It could be stored in a vacuum where it would be prevented from touching anything, including the sides of its own tank, by powerful magnetic fields.

Risky business. A momentary brownout or weakening of the containment field and ka-*plooie.* We're probably a few centuries away from enlisting volunteers for the ride.

And we may not have to. If we could glimpse the technology of the future it would seem indistinguishable from magic. Someone from the last century transported to our modern age might accept flying machines and telephones but would have serious difficulty with holograms. Witchcraft, surely. What would we think if we could peek at the science of the twenty-second century?

So we need not strain to figure out how to reach the stars. The Galaxy's distant diamonds will fall into place as naturally as a toddler's first steps when the time is ripe, their means of attainment self-evident.

It's not yet our time for star travel, nor for planetary colonization. Ours is the age of robotics, of computer advances, of data from unmanned craft beamed from other worlds to our minds and to our spirits.

Quite enough to keep us busy for now and, perhaps, to keep our mischievous hands from putting fingerprints on places that are still undisturbed. And yes, the very unattainability of the stars has its own beauty. We're like children in a fabulous museum of sparkling gemstones, told that we can look but not touch.

To which we tell ourselves, "Yes, but when I'm older . . ."

Appendix 1
The Ten Telescope Secrets

What greater pleasure than to stroll the universe's mysterious boulevards? The starry path can take us no farther than an easy chair, where a world of magazines and books distills the fascinating flood of knowledge pouring in from observatories and from machines in outer space. Or we can actively reach out to the stars through the impressive instruments most local astronomy clubs regularly set up, free, for public use. A further option exercised by millions around the world is ownership of a telescope or astronomical binoculars.

Appendix 2 provides basic guidance on selecting and buying a telescope or binoculars. If you already own such an instrument, considerable time and energy can be saved with what I call the ten telescope secrets. And while they may be "secrets" to the novice, every telescope owner will learn them the hard way over time. One may as well hit the ground running, by avoiding these pitfalls from the outset.

1. Never point a telescope through a window, open or closed, unless you're spying on neighbors. The currents of air from an open window will seriously degrade the image. A closed window is even worse: Window glass is of inadequate optical quality.

2. Do not set up a telescope on a deck, a porch, or any other wooden structure. Vibrations from your body, though they may seem imperceptible, are amplified by frame construction to make the image vibrate. Telescopes must be used outdoors only on an earthy foundation, either natural or cement. Lawns are ideal.

3. Always start with low power, and keep it low whenever possible. Stars will appear crisper and images brighter and sharper with your low-power eyepiece. This, oddly enough, will be the one with the highest number written on it: A 40-millimeter eyepiece yields one-fourth the magnification of a 10 millimeter. You can determine the eyepiece's power by dividing its number into the telescope's focal length. (How to find *that?* It's probably written on the tube itself, as "f =" followed, typically, by a number between 700 and 1,500 millimeters.)

A further advantage of low power, unless your instrument is motorized and polar-aligned, is that objects will stay in the field longer. As a general rule, 50 × is far more useful than 300 ×.

4. Do not clean your telescope. At least, don't clean it often. If it's a reflector, a dusty-looking mirror may bother you, but will not degrade the image very much. Frequent cleaning, on the other hand, *will* harm it. And do not clean it at all if you don't know how. Rubbing a telescope mirror with a lens-cleaning cloth is not much better than blasting it with a shotgun.

M13, a globular cluster of stars beyond the constellation Hercules. It is one of the finest objects seen in six-inch and larger telescopes.

5. Learn to find a few new celestial sights each year. Most telescope users can locate no more than two objects, usually the moon and Jupiter. Explore on your own or with friends, or see Appendix 4 for a subjective list of the most impressive things seen through a medium-size amateur telescope.

6. Do not expect a planet to look impressive the moment you point your instrument its way. Atmospheric turbulence is almost always present, and varies continuously. Experienced observers stay at the eyepiece, often for hours, alert for the moment when the air suddenly steadies to reveal exquisite detail.

7. Plan your observing itinerary around existing conditions. Do you have clear, transparent skies? Or do you instead have *steady* skies, when the stars aren't twinkling? Is it a little hazy? Is moonlight present or absent? These varying conditions call for entirely different targets.

Steady skies, and nights with bright moonlight, tell you to view planets, the moon, and double stars. A little bit of haze or fog not only won't hurt them, but often translates into superb lunar or planetary viewing. But forget anything else.

Conversely, dark, clear, moonless conditions tell you to focus on deep-space objects such as nebulas, star clusters, and galaxies. These objects demand the transparent air whose signature is a sky flooded with myriad stars. Because such deep-space objects are nebulous by nature and have no fine detail to be resolved, they're forgiving of the shimmering that so wrecks the images of planets.

8. Use peripherals to make life easier. Gloves, scarf, and long johns in cold weather. Flashlight with red lens for reading maps or finding things. Hairdryer if someone breathes on the lens. Maybe a pirate patch so you can keep both eyes open at the eyepiece.

9. Pass binoculars to guests by first putting the strap over their heads. That way if they drop the instrument, they don't drop it. Dropped binoculars rarely suffer broken or cracked lenses. What happens is that they go out of alignment, so that each component points in a slightly different direction, producing double images. When this occurs, throw them away. The cost of repair almost always exceeds the price of a new instrument.

10. Instruct your guests in telescope procedure. Tell them: Don't touch. People routinely grab or lean on telescopes as they look through them, causing them to point somewhere else. "Don't touch" also applies to the surfaces of lenses and especially mirrors (and maybe even yourself, if the wrong person starts getting carried away by the moon's enchantment).

If you're in an area where a passing car's headlights or other bright lights may suddenly intrude, another basic procedure is not to look their way. Night vision takes at least ten minutes to acquire; a single headlight resets the clock to zero. Do not expect guests to know these things. You must tell them.

But make no mistake. You do not need a telescope for exploring many of the wonders of the night sky. And while it is a valid and venerable procedure merely to let yourself be swept mindlessly by the grandeur of the stars, a precious tool is knowledge, knowledge that can come easily with a subscription to a good astronomy or science magazine such as *Sky and Telescope* or *Discover.* They'll keep you posted on upcoming celestial events as well as the continuing discoveries of space science.

Meantime, you need only take time from earthly concerns to drink in the rapture of the deepest deep, and compile your own list of the most amazing things in the universe.

Appendix 2
Buying a Telescope

You do not need a telescope to explore the universe. This book overflows with examples of celestial treasures awaiting the nature lover who is armed with nothing more than the naked eye.

Yet, undeniably, there are marvels such as Jupiter, Saturn, and the great globular clusters of hundreds of thousands of distant suns whose essential glory dawns solely through the telescope.

Unfortunately, the beginner often gropes helplessly along the labyrinthine corridors of conflicting claims by telescope manufacturers,

Details on Jupiter seen through an amateur telescope

and the ancient question of how much is proper to spend on a hobbyist instrument. While a thorough treatment of the topic lies beyond the space of a mere appendix, the fundamentals of selecting a telescope are not difficult.

Sadly, the same scam has been successfully and continuously pulled on the public during the four centuries since the telescope's invention. The scenario goes like this:

You walk into the department store or camera shop to buy a small telescope and, of course, there are several models. Forget receiving advice from behind the counter: Unless you've lucked into an astronomy buff, you and the clerk are probably well matched in your ignorance of optics. So, how to decide?

The manufacturers know how you will choose. You will select the most "powerful" instrument for the money. And since *powerful* has the word *power* in it, you can't be blamed for assuming that the highest power or magnification means the best telescope.

To be fair, some manufacturers would love to sit you down and instruct you in the ABC's of optics. But as things stand, they know you do not realize that all telescopes have a primary lens or mirror that gathers light, and that the larger this is, the more powerful the telescope will be. In other words, if you looked into the telescope's wrong end you'd see a big lens (or down the tube a curved mirror); it's the diameter of this so-called objective lens or mirror that determines the instrument's ability to reveal the secrets of the night.

But you don't know this. If you did, you'd grasp that the *thickness* of the telescope's tube rather than its length or magnification is a dead giveaway to its real worth. The *fatter* the telescope, the better—because it is supporting a larger light-gathering lens or mirror. Perhaps you'd then ask the clerk, "What's the diameter of the telescope's objective?" You'd grasp that, all else being equal, an "eight-inch" telescope is better than a "four-inch," and you'd be correctly speaking the language of size rather than power.

Instead, many companies routinely include an eyepiece yielding a wildly exaggerated magnification, since almost any "power" can be hooked up to any telescope. Small department-store instruments cannot produce a clear and bright image at any power above 100 × or so. But eyepieces cost manufacturers only a few dollars each, and if they don't include a 300 × or even 500 × eyepiece (so they can proclaim in big letters 500 POWER TELESCOPE), they know their competitors will.

So there you are, looking blankly at the instruments on display. If they're the same price, but one says 120 × and the other boasts 500 ×, your response will be as predictable as the sunrise: "Here's a more powerful telescope for the same price. I'll take it!"

To your regret, you'll quickly find the images through the high-power eyepiece to be uselessly dim and fantastically blurry. You've been lured into an "all you can eat" establishment only to be served a mountain of mush.

So the rule is simple: *Ignore all claims about power (magnification).* Just be sure a $130 **refractor** has at least a 60-millimeter objective lens, and a **reflector** a primary mirror at least 4¼ inches in diameter. Refractors (the kind you look "through") are generally hardy and hassle free, while reflectors (which you face sideways) deliver brighter images for the dollar but demand periodic alignment of the optics. No great job, but if you're the type who refuses to fiddle with things, then it's ix-nay on reflectors.

In either case, the instrument will come with a simple **altazimuth** mount

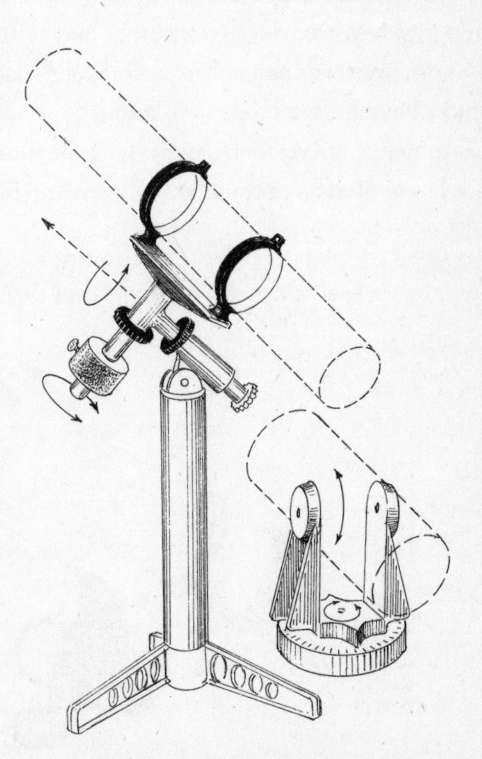

Left: *an equatorial mount. The dashed arrow is aimed at the celestial pole, allowing easy tracking of the target.* Right: *an altazimuth mount. It simply swivels up and down, and rotates like a lazy Susan.*

(a wonderfully fancy way of saying the telescope swivels up and down and side to side) or else the more expensive **equatorial** mounting, which allows objects to be tracked more easily as they cross the sky. The equatorial system looks satisfyingly high tech but adds $75 to $150 to the cost, and is useful only if you're going to take the time to polar-align it properly for each session. If you're not inclined toward that small task, save the money and get the cheaper mounting.

If you can afford it, spring for the $500 to $900 for a 6-inch or even 8-inch reflector telescope with a motor drive that tracks the stars. Such instruments can supply a lifetime of enjoyment. And they use the modern eyepieces of 1¼ inches in diameter, a major improvement over the generally pathetic, restricted-field 0.965-inch barrels of cheaper instruments. (Bring a tape measure to the store!) The better telescopes generally come through dealers or directly from the manufacturer. Check the ads in astronomy magazines.

Usually the beginner turns to the same handful of celestial objects time and again, and the instrument is soon sold or stored in the attic. A minority, however, develop a serious and continuing interest in astronomy and discover ever newer (but increasingly esoteric) objects to observe. If you're buying a telescope for a young person, start off with an inexpensive instrument until you can determine if the interest is just a passing fancy or if it has ignited an enthusiasm to learn and do more.

Two typical reflecting telescopes: a Newtonian reflector (top) *and a Schmidt-Cassegrain. The fatter the telescope, the more powerful it will be.*

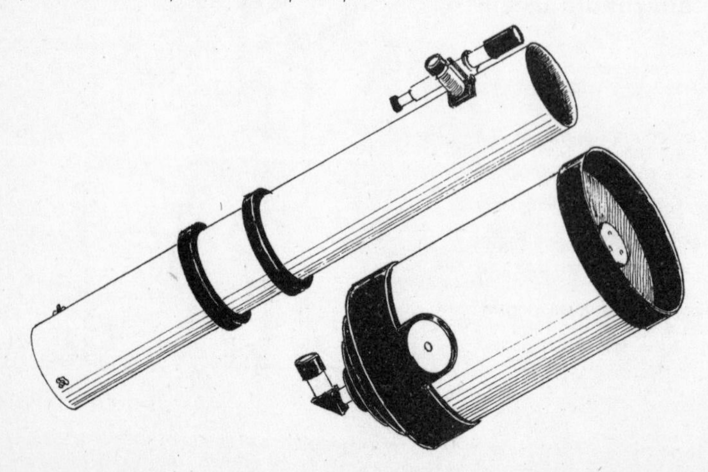

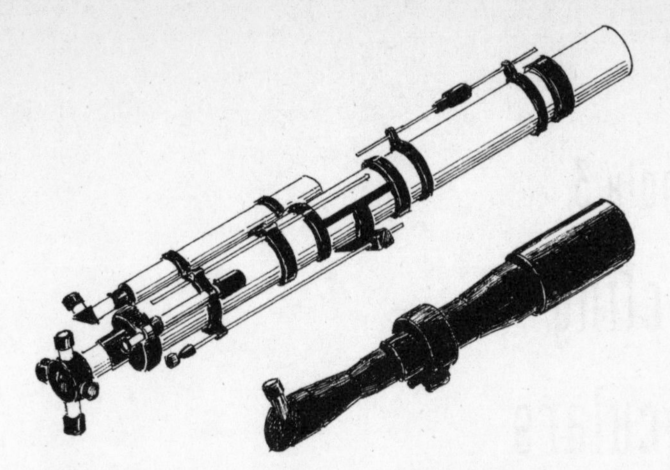

Two typical refracting telescopes. Hassle free, this is the type of telescope you look "through."

Do not overspend. Not much is worthwhile in the middle price range. Below $150 one finds the 60-millimeter altazimuth refractors that are such ideal and rugged beginner instruments. Above $500 start the 6-inch, 8-inch, and larger reflectors with tracking motors ("clock drives") and the portable Schmidt-Cassegrain types. The latter are particularly useful if the telescope will also be used terrestrially, such as for bird watching or taking photographs for blackmail.

As for choosing among the three main types, a reflector will give you the best image for the dollar, at the expense of being a bit unwieldy and requiring occasional mirror collimation, which is a simple job. Try to get one with a focal ratio of f/6 or higher; the common f/4 models are extremely sensitive to mirror misalignment.

A refractor is a great no-fuss instrument. It has no peer in the 60-millimeter to 80-millimeter class, but gets expensive when its objective lens is larger than 3 or 4 inches.

The Schmidt-Cassegrain is a short-tube, portable instrument whose greatest advantage is portability. If you'll also be using it for nature watching and if you live in a light-polluted city, it's a strong contender.

A telescope, however, is not the ideal instrument for all observations. Many celestial targets are best seen through ordinary binoculars. If you don't already own a pair, Appendix 3 may be helpful.

Appendix 3
Selecting Binoculars

The biggest bargain in the realms of nature and astronomy is a pair of simple binoculars. Crisp, bright, astonishingly beautiful images are to be had for less than one hundred dollars. And contrary to popular thought, an image seen through binoculars is superior, not inferior, to that of a telescope, for the simple reason that the relaxed use of both eyes yields a more enjoyable impression as well as a sense of 3-D.

Moreover, many celestial objects look far better through binoculars than through the largest telescopes on Earth. That's because low magnification and a wide field are the requirements to properly view sprawling objects such as the larger galactic star clusters.

Because good binoculars are invaluable and inexpensive, no observer or nature lover should be without them. Buying the right pair, however, requires a bit of knowledge and tenacity; you must be willing to dive into a sea of numerals.

Specifications for binoculars contain two numbers separated by an ×—as in 7 × 50. The first numeral is the magnification, and should be between 7 and 11: Anything higher cannot be steadily hand held and requires a tripod. Seven or 8 power is close to ideal.

The second number is more important. It's the diameter of the objective lenses in millimeters. This should never be less than 35, eliminating from celestial consideration the miniature "shirt pocket" models, which have lenses between 20 and 25 millimeters. Some unscrupulous companies imply in their ads that, like computers, binoculars have been successfully miniaturized by modern technology. In reality, the small models merely use tiny lenses. Nothing high tech about this at all—and, as you'd expect, these instruments yield much fainter images in low-light conditions.

Further insight about image brightness can be decoded from the specifications. Dividing the second number by the first yields the **exit pupil,** the diameter of the shaft of light that enters the pupil of your eye. For example, a 7 × 35 has a 5-millimeter exit pupil. Reasonably young and healthy eyes dilate to about 7 millimeters in dark skies, so a model yielding a 7-millimeter exit pupil (such as 7 × 50) gives the brightest image for night work. Such instruments are often called **marine binoculars** or **night glasses** because all nocturnal targets, including a dark horizon at sea, are maximally clear and bright through them.

Conversely, an exit pupil below about 3.5 renders the image too dim for night work, which is why shirt-pocket 8 × 21 models are limited to daytime use or well-lit sports arenas.

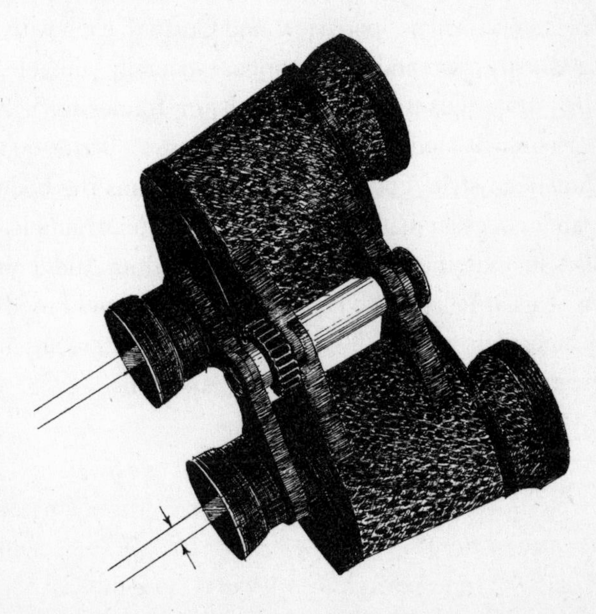

The diameter of the light beam entering your eye is called the exit pupil. The bigger this is, the brighter the image.

The only potential drawback of large 50-millimeter binoculars is that they brighten things so much that in urban or suburban settings the sky's milkiness is exaggerated. The resulting view of stars against a gray instead of black background is an aesthetic no-no. So if your skies aren't pure and dark, 7 × 35's (or 8 × 40's, or anything else with an exit pupil of 5 or so) are probably best, and are excellent all-purpose terrestrial "glasses" as well.

Try out any binoculars first, in the store, by going down the following list:

Focus carefully on a distant object and note how much of the field of view stays entirely in focus. Most binoculars, especially the "wide field" variety, will display blurriness around the edges, and this is all right. But be sure that at least the central three quarters of the field is razor-sharp.

Next, focus on something with a high brightness contrast, such as the moon in the night sky, or a white store sign against a dark background. Inspect the image for fringes of false color (typically purple, orange, or yellow). Again, some degree of chromatic aberration is normal. Make sure it isn't excessive or bothersome.

Try several pairs of binoculars. The first may look fine, but comparing the images will let you detect distinct differences in sharpness and contrast. When the image looks better, it *is* better.

Brand by itself doesn't tell the full story, because manufacturers typically produce a range of models. Bushnell, for instance, makes the Banner, Spectator, Ensign, Sportsview, and Custom, each with successively better construction and higher price. Generally you get what you pay for. High-price pluses include magnesium frames for light weight, rubber ("armored") models for waterproof use, better-quality lenses, and "American-style" construction, which means the body is of one-piece, seamless design that holds collimation better. Examples of superb binoculars include the Bausch and Lomb Custom Audubon models. During my years at Yellowstone Park, several rangers and I tried out endless pairs of binoculars brought by visitors, and while there are many fine ones on the market, we fell in love with the Audubons.

Appendix 4

The Twenty Most Impressive Telescopic Targets

Summer

OBJECT AND DESCRIPTION	MAGNIFICATION*	REQUIRED SKY CONDITIONS
Ring Nebula M57: ghostly doughnut shining with "forbidden radiation"	M	very clear and dark
Albireo: gorgeous double star, components blue and deep yellow	VL, L	all
M11: lovely open (galactic) cluster	VL	clear
Saturn: best planet	M, H	steady; haze okay, bright moonlight okay
Omega Nebula: summer's best nebula	L	clear and dark

*VL = very low: < 60 ×. L = low: 60–100 ×. M = medium: 100–260 ×. H = high: > 260 ×.

Autumn

OBJECT AND DESCRIPTION	MAGNIFICATION	REQUIRED SKY CONDITIONS
Double Cluster in Perseus: finest open cluster in the heavens	VL	dark
TX Piscium: deepest-red star	any	all
Gamma Delphini: colorful double star	L, M	all
M27: planetary nebula; greenish	M	clear and dark

Winter

OBJECT AND DESCRIPTION	MAGNIFICATION	REQUIRED SKY CONDITIONS
M42: Orion Nebula; the sky's best	VL, L, M	clear and dark
Sirius, spectroscopically	L	all
M35 and M46: two nice open clusters	L	clear and dark
R Leporis: deep-red star	L, M	all

Spring

OBJECT AND DESCRIPTION	MAGNIFICATION	REQUIRED SKY CONDITIONS
M13: great globular in Hercules; may be the finest telescopic object	M	clear, dark
NGC 4565: superb edge-on galaxy	L, M	clear, dark
M51: Superb face-on "whirlpool" galaxy; shows traces of spirality!	L, M	clear, dark
M82: interesting exploding galaxy	L, M	clear, dark

OBJECT AND DESCRIPTION	MAGNIFICATION	REQUIRED SKY CONDITIONS
Jupiter (low power for its moons; medium or high for "surface" detail; steady air a must)	L, M	steady, but haze or moon okay
Cor Caroli: oddly tinted binary	L, M	all
NGC 6543: interesting planetary nebula; very green	M	all

References for the Home Observatory

Abell, George. *Exploration of the Universe.* 6th edition. Philadelphia: W. B. Saunders, 1993. The best textbook. Answers all questions of a nonobservational nature—black holes, Einstein, planets, and so on. With beautiful photographs.

Burnham, Robert. *Burnham's Celestial Handbook.* 3 vols. Mineola, NY: Dover, 1966. An indispensable, beautiful, exhaustive survey of the entire sky. Science, myth, legends, and lore. A must.

Canadian Astronomical Society. *Observer's Handbook.* Published yearly. Lists the current coordinates of all objects of interest, and interesting celestial events for that year. Available through Sky Publishing.

Ridpath, Ian, ed. *Norton's 2000.0 Star Atlas and Reference Handbook.* 18th edition. New York: Halsted, 1989. A must for any observatory. Clearest sky maps.

Sky Catalog 2000. Cambridge, MA: Sky Publishing, 1982. A data reference of all the stars. For the advanced amateur.

Tirion, Will. *Sky Atlas 2000.0.* Cambridge, MA: Sky Publishing, 1981. A must.

Sky Publishing is a good source for useful astronomical publications. Telephone number: (800) 253-0245.

Index

Page numbers *in italics* refer to illustrations and captions.